Automatic Detection Algorithms of Oil Spill in Radar Images

Maged Marghany

Director Institute of Geospatial Applications
Faculty Geospatial and Real Estate
Geomatika University College
Kuala Lumpur, Malaysia

CRC Press
Taylor & Francis Group
Boca Raton   London   New York

CRC Press is an imprint of the
Taylor & Francis Group, an **informa** business

A SCIENCE PUBLISHERS BOOK

Cover illustrations provided by the author of the book, Dr. Maged Marghany

CRC Press
Taylor & Francis Group
6000 Broken Sound Parkway NW, Suite 300
Boca Raton, FL 33487-2742

First issued in paperback 2021

© 2020 by Taylor & Francis Group, LLC
CRC Press is an imprint of Taylor & Francis Group, an Informa business

No claim to original U.S. Government works

Printed on acid-free paper
Version Date: 20190819

ISBN-13: 978-0-367-14660-3 (hbk)
ISBN-13: 978-1-03-208886-0 (pbk)

Visit the Taylor & Francis Web site at
http://www.taylorandfrancis.com

and the CRC Press Web site at
http://www.crcpress.com

Dedicated to
My Mother Faridah
and
Nikola Tesla and Richard Feynman who taught me that
real works of professors last forever while fake works of
professors do not.

Preface

These days, the new generation of Synthetic Aperture Radar (SAR) sensors necessitates the growth of new procedures for trustworthy data processing and information abstraction. Yet no work has bridged the gap between modern physics and complete understanding of microwave theories. Indeed, microwave remote sensing theories and techniques are based on modern physics. Moreover, there is also a great gap between modern applied physics and SAR data image processing. In this view, the main image processing techniques of SAR data is restricted by speckle filter procedures, classical edge detection tools and mathematical arithmetic operation, which, especially applies to SAR polarimetry data processing. The growth in SAR data processing, mainly for automatic detection of oil spill shows slow development in term of algorithms. The majority of automatic detection oil spill algorithms to SAR data are restricted conventional image processing tools, which are based on classical image processing techniques such as image segmentation and classical learning machine algorithm, for instance, artificial neural networks and support vector machine algorithms. Yet these procedures are not able to simulate the trajectory movement of an oil spill or to forecast its path from a single SAR data or multiSAR data.

Recently, SAR data have proven the great potential for monitoring and tracking the Deepwater Horizon oil spill disasters. However, the SAR data are only used for automatic detection of an oil spill without demonstrating the gradient variation across the spill-covered water, i.e., oil spill spreading in SAR data. In this view, Synthetic Aperture Radar Imaging Mechanism for Oil Spills delivers the critical tool needed to understand the latest technology in radar imaging of oil spills, particularly microwave radar as the main cradle to monitor and precisely detect marine oil pollution. To this end, modern physics such as quantum mechanics must be involved in microwave theories and their data processing. This might lead to a new era of the quantum microwave, quantum computing and quantum image processing.

The aim of this book is to viaduct the mismatch between modern physics, quantum mechanism and applications of radar imaging and automatic detection algorithm of the oil spill. This book is divided into mechanical details to assist in the potentiality of synthetic aperture radar (SAR) and key approaches to be depleted to extricate the worth-novel information crucial, for instance, location, size, perimeter and chemical details of the oil spill from SAR measurements. Rounding out with practical simulation rajectory movement of oil spills using radar images, Synthetic Aperture Radar Imaging Mechanism and processing for Oil Spills conveys an operative novel stoolpigeon of modern machinery and is used by present day oil and marine pollution engineers.

This book delivers a comprehensive understanding of Maxwell's equations. In fact, these equations are the keystone to understanding the speculations of the microwave remote sensing. Truly, the common microwave theories are imperative to comprehend the physics of Maxwell's equations. However, the majority of graduates of geospatail and microwave specialists no longer correlate between microwave remote sensing and Maxwell's equations. In this circumstance, the foremost standing is how to run a variety of SAR filter techniques to minimize speckles. In general, mapping is a general output product based on microwave data without grasping the key theory behind microwave remote sensing. Consequently, the copiously comprehensive Maxwell's equations are described in Chapter 1.

It is impossible to deal with photons as a basic of microwave remote sensing without understanding their mechanical behavior. Chapter 2 reveals the quantum mechanics theories that explain in depth the behavior of photons as a core of electromagnetic as a function of Maxwell's equations. In continuation with Chapter 2, Chapter 3 demonstrates the novel theory of Josephson junctions to understand the behavior of microwave photons. While, Chapter 4 describes the scattering theory from the point view of the quantum mechanics. In this view, the quantum radar theories can be established. State-of-the-Art, professor Dr. Marco Lanzagorta is the only pioneering scientist who associated quantum mechanism to Maxwell's equations and radar imaging techniques. The speculative of the quantum radar mechanisms, consequently, is initially delivered to originate the novel decoherence quantum theory of oil spill imaging in Synthetic Aperture Radar in Chapter 5. However, the conventional principle of Synthetic Aperture Radar is also addressed in Chapter 6 to fill the gap between quantum radar theories and conventional SAR techniques. Moreover, the relativity theories are also involved in understanding the radar image mechanism. Chapter 8 also reveals the quantization of oil spill imaging mechanism in synthetic aperture radar.

The book delivers conventional algorithms for automatic detection of oil spills. For instance, texture algorithms and conventional machine learning, i.e., Mahalanobis, neural network and fractal algorithms are described in Chapters 9, 10, and 11, respectively. However, Chapter 9 introduces a novel algorithm, which is based on the quantum entropy for automatic detection of oil spill.

The book also delivers a new approach for automatic detection of the oil spill by implementing a new approach of Quantum-dot Cellular Automata, which has never been implemented before this book. A more advanced study and a new work of quantum multiobjective algorithm is presented in Chapter 13 for automatic detection of oil spill spreading in full polarimetric SAR data. The last chapter introduces a novel approach for simulation and forecasting oil spill trajectory movements based on quantum Hopfield algorithm. This algorithm is implemented with multiSAR satellite data. The quantum Hopfield achieves automatic detection of oil spill spreading and forecasts the oil spill trajectory movement over different time of SAR data acquision.

Prof. Dr. Maged Marghany

Microwave Remote Sensing Expert
Director Institute of Geospatial Applications
Faculty Geospatial and Real Estate,
Geomatika University College,
Taman Setiawangsa, 54200,
Kuala Lumpur, WP Kuala Lumpur,
Malaysia

Contents

Microwave Remote Sensing Based on Maxwell Equations

Maxwell's equations are the cornerstone to understand the principles of the microwave remote sensing. The common microwave concepts are imperative to comprehend the physics of Maxwell's equations. Consequently, the large comprehensive Maxwell's equations are discussed in this chapter. The chapter gives a scientific clarification of Maxwell's equations and their mathematical and physical characteristics. In these regards, all electromagnetic phenomena can be said to follow from Maxwell's equations.

However the majority of microwave specialists do no longer correlate between microwave remote sensing and Maxwell's equations. In this circumstance, the foremost standing is how to run a variety of filter analyses to microwave data to minimize noise. In third world countries, the majority of microwave scientists are interested in mapping without understanding the main concept beyond microwave remote sensing. The common question that arises is, where did these equations come from?

1.1 Maxwell's Equations

Throughout the 1860s, the Scottish physicist James Clerk Maxwell revealed an amalgamated speculative narrative of entirely electrical and magnetic spectacles. Maxwell's equations are one of the pronounced successes of the genius philosophy. These equations are the cornerstone of the relativity theory. Indeed, Einstein far along designated Maxwell's theory as 'the greatest insightful and the furthermost rewarding that physics has inspired later, Newton's era.' The equations are wonderfully developed and worth showing to the readers. In contrary to with those who deliberated that microwave remote sensing must be grasped without mathematical equations and those mathematical equations are a reason beyond the reduction of the number of rummage sales of prevalent scientific manuscripts.

Maxwell's equations are formulated as [1]:

$$div\ D(=\nabla.D) = \rho \tag{1.1}$$

$$div\ B(=\nabla.B) = 0 \tag{1.2}$$

$$curl\ E(=\nabla x E) = -\frac{\partial B}{\partial t} \tag{1.3}$$

$$curl\ H(=\nabla x H) = \frac{\partial D}{\partial t} + J \tag{1.4}$$

The *Es* and *Hs* stand for electric (Fig. 1.1), and magnetic fields (Fig. 1.2), respectively. Consequently, the **electric field** is termed as the **electric** energy per unit charge. The path of the **field** is considered as the route of the energy it **would** make use of a positive test charge. In this understanding, the **electric field** is centrifugally away from a positive charge and centrifugal toward a negative point charge. In this view, the elementary construction modules of Maxwell's explanation of the electric and magnetic phenomena. Furthermore, ρ is the charge density and J is the current density. The electric displacement and magnetic induction are represented by D; and B, respectively [2–4].

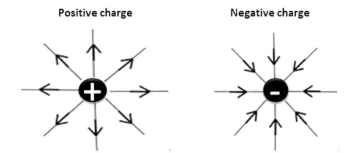

Figure 1.1. Electric field.

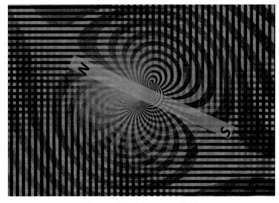

Figure 1.2. Magnetic field.

Let us assume that there are constitutive relationships between D to E, B to H; and J to E; and the medium to be isotropic. In other words, D is on the same route as E. In this circumstance, constitutive relationships are expressed as:

$$D = \varepsilon E \tag{1.5}$$

$$B = \mu H \tag{1.6}$$

and

$$J = \sigma E \tag{1.7}$$

here, ε, μ, and σ are termed as the dielectric permittivity, magnetic permeability and the conductivity of the medium, respectively. Consequently, both σ and ρ equal zero in the circumstance of a charge free dielectric; therefore, J also equals zero [5]. These constitutive relationships precede to abridge Maxwell's equations as follows:

$$\nabla.D=0 \tag{1.8}$$

$$\nabla.H=0 \tag{1.9}$$

$$\Delta x(\nabla x E) = -\mu \frac{\partial}{\partial t} \nabla x H = -\varepsilon\mu \frac{\partial^2 E}{\partial t^2} \tag{1.10}$$

and

$$\nabla x H = \varepsilon \frac{\partial E}{\partial t} \tag{1.11}$$

If we substitute H in terms of E in Equation 1.11 and use the vector identity, the following equation:

$$\nabla x(\nabla x E) = \nabla(\nabla.E) - \nabla^2 E \tag{1.12}$$

Equation 1.12 can be cast as:

$$\nabla(\nabla.E) - \nabla^2 E = -\varepsilon\mu \frac{\partial^2 E}{\partial t^2} \tag{1.13}$$

Under the circumstance of $\Delta.E = 0$, which represents a homogeneous medium, Equation 1.13 can be expressed as:

$$\nabla^2 E = -\varepsilon\mu \frac{\partial^2 E}{\partial t^2} \tag{1.14}$$

Likewise, if curl is implemented to Equation 1.11, then H satisfies the following equation:

$$\nabla^2 H = -\varepsilon\mu \frac{\partial^2 H}{\partial t^2} \tag{1.15}$$

1.2 Simple Wave Equation Based on Maxwell's Equations

Both Equations 1.14 and 1.15 are known as the three-dimensional wave equation for each Cartesian components of E and H, which are satisfied with the scalar wave equation as follows [6–10]:

$$\nabla^2\psi = v^{-2}\frac{\partial^2\psi}{\partial t^2} \tag{1.16}$$

being

$$v = \left[\sqrt{\varepsilon\mu}\right]^{-1} \tag{1.16.1}$$

Equation 1.16 demonstrates the wave propagation with velocity v in free space, which is estimated via [14–17]:

$$\varepsilon = 8.8542x10^{-12}\frac{C^2}{Nm^2} \tag{1.17}$$

$$\mu = 4\pi x10^{-7}\,Ns^2C^{-2} \tag{1.18}$$

In this regard, the speed of electromagnetic wave propagation is computed via:

$$v = c = \left[\sqrt{8.8542x10^{-12}x4\pi x10^{-7}}\right]^{-1} = 2.99794x10^8\,ms^{-1} \tag{1.19}$$

Equation 1.19 demonstrates the speed of the light, i.e., $3x10^8ms^{-1}$, which conveys that light must be an electromagnetic wave. This is why Maxwell's equations remain all the time one of the greatest triumphs of human intellectual endeavor [11–15]. In this view, the simplest wave propagation in the positive z-direction is expressed as:

$$\psi = Ae^{i(\omega t - kz)} \tag{1.20}$$

where A is wave amplitude, ω is a frequency, and k is wave number. The simple mathematical relationship between frequency and wave number can be written as [15–18]:

$$k = \omega v^{-1} \tag{1.21}$$

In Equation 1.16.1, the wave number k can have a relationship with the root square of dielectric permittivity (ε) and magnetic permeability (μ) and can be written as:

$$k = \omega\sqrt{\varepsilon\mu} = \omega nc^{-1} \tag{1.22}$$

Equation 1.22 shows that the wave number is a function of frequency and the inverse of the speed of light c.

In general, Maxwell's equations deliver a set of expressions for electric and magnetic fields everywhere in space postulated that all charges and current sources are delineated. Consequently, they symbolize one of the most sophisticated and abridged ways to declare the fundamentals of electricity and magnetism [10–14]. Additionally, they pronounce the association between the electric and magnetic fields and sources in the medium. They also express a high level of mathematical sophistication owing to their concise statement. Explaining masses of phenomena in such few simple sets of equations was a foremost exploit that led to Einstein valuing Maxwell's accomplishment on equivalence with that of Newton. Einstein grasped the Maxwell's idea and combined them into his relativity theories [1,6,10,18]. In Einstein's equations, magnetism and electricity were indexes of the same thing comprehended by viewers in diverse frames of reference; an electric field in one moving frame would be seen as a magnetic field in another [11–14]. Conceivably, it was Einstein then who eventually machinated that electric and magnetic fields are exceptionally one and the same thing.

1.3 Solution of Electromagnetic Waves in a Homogenous Dielectric

Electromagnetic waves involve electric and magnetic fields, which can travel through a vacuum starved of any an associated medium, and do not contain moving charges or currents. The presence of electromagnetic waves was originally established in 1888 by the German physicist Heinrich Hertz (1857–1894). Hertz exploited an RLC circuit that generated a current in an inductor that drove a spark gap. In this view, an RLC circuit is an electrical circuit consisting of a resistor (R), an inductor (L), and a capacitor (C), connected in series or in parallel with voltage V. The name of the circuit is derived from the letters that are used to denote the constituent components of this circuit, where the sequence of the components may vary from RLC (Fig. 1.3) [17].

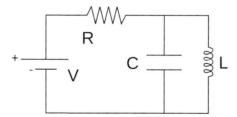

Figure 1.3. Basic of RLC circuit.

A flash cavity involves two electrodes that, when an impending difference is smeared through both electrodes, yield a flash by electrifying the gas between them. In this circumstance, Hertz positioned a loop and a small catalyst gap numerous meters away from each other. He perceived that flashes were generated in the isolated loop in a form that interrelated with the electromagnetic oscillations in the primary RLC circuit. Accordingly, electromagnetic waves were capable to propagate across space-starved of any medium to sustenance them. In this regard, the rudimentary unit of oscillation, cycles per second, was entitled as the Hertz (Hz) in his honor [16–18].

The common 3-D plane electromagnetic wave can be mathematically given by:

$$\psi = Ae^{i(\omega t - \vec{K}.\vec{r})} \tag{1.23}$$

Substitute Equation 1.23 into 1.20 we would obtain that

$$k_x^2 + k_y^2 + k_z^2 = k^2 = \omega^2 n^2 c^{-2} = \frac{\omega^2}{v^2} \qquad (1.24)$$

In this view, the electric and magnetic fields of plane wave propagation in the direction $\vec{K}$ are formulated as [1,5]:

$$\vec{E} = \vec{E}_0 e^{i(\omega t - \vec{K}.\vec{r})} \qquad (1.25)$$

and

$$\vec{H} = \vec{H}_0 e^{i(\omega t - \vec{K}.\vec{r})} \qquad (1.26)$$

Equations 1.25 and 1.26 explore both electric and magnetic fields as space and time-independent vectors. Conversely, both fields are complex. In this context, the equation $\nabla.\vec{E} = 0$ delivers [1,5]:

$$-i(k_x E_{0x} + k_y E_{0y} + k_z E_{0z}) = 0 \qquad (1.27)$$

Equation 1.27 can be simplified as;

$$\vec{K}.\vec{E} = 0 \qquad (1.28)$$

Similarly, the mathematical equation $\nabla.\vec{H} = 0$ can be expressed in the form of the plane wave propagation in the direction $\vec{K}$ as follows [1,17]:

$$\vec{K}.\vec{H} = 0 \qquad (1.29)$$

Equations 1.28 and 1.29 demonstrate that both electric and magnetic fields are at right angles of the wave propagation in the direction of $\vec{K}$. These can be mathematically written as:

$$\vec{H} = \vec{K}x\vec{E}(\omega\mu)^{-1} \qquad (1.30)$$

$$\vec{E} = \vec{K}x\vec{H}(\omega\varepsilon)^{-1} \qquad (1.31)$$

In the circumstance of $\vec{E} x \vec{H}$ in the direction of propagation z, then the electric vector to be along the x-axis and the magnetic vector would be along the y-axis. With this understanding, the mathematical description of the propagation of both fields is given by [14–17]:

$$\vec{E} = \vec{x}E_0 e^{i(\omega t - kz)} \qquad (1.32)$$

$$\vec{H} = \vec{y}H_0 e^{i(\omega t - kz)} \qquad (1.33)$$

In this regard, the actual electric and magnetic fields are the real part of the exponentials appearing as the electromagnetic wave of Equations 1.32 and 1.33. In this view, the propagation of an electromagnetic wave can express mathematically in the form of a cosine wave as:

$$\vec{E} = \vec{x}E_0 \cos(\omega t - kz) \qquad (1.34)$$

$$\vec{H} = \vec{y}H_0 \cos(\omega t - kz) \qquad (1.35)$$

In this understanding, a cosine wave is a signal waveform with a form indistinguishable to that of a sine waveform, excluding a piece point on the cosine wave transpires exactly 1/4 cycle prior than the conforming time on the sine waveform (Fig. 1.4) [17].

Furthermore, the linear polarization of the plane wave can be delivered by both Equations 1.34 and 1.35 and the directions x and y must be perpendicular to each other. In other words, the x-axis represents the electric vector while the y-axis represents the magnetic vector (Fig. 1.5).

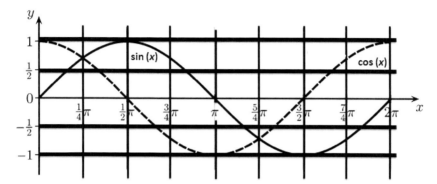

Figure 1.4. Sine and cosine waveforms.

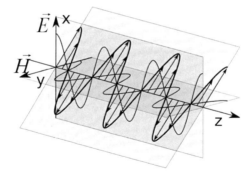

Figure 1.5. Linear polarized electromagnetic wave.

1.4 Electromagnetic Wave Characteristics Based on Maxwell's Equations

Maxwell's equations explain that both electric and magnetic fields are considered as the two vectors, which are at a right angle to the direction of the propagation (Fig. 1.6). The vectors $\vec{E}$ and $\vec{H}$ are perpendicular to each other, such that the direction of the vector $\vec{E} x \vec{H}$ is along the route of the electromagnetic wave propagation z. Hence, if the path of the transmission is along the z-axis and if $\vec{E}$ is supposed to indicate the x-axis direction, formerly $\vec{H}$ will direct in the y-direction [1,12,15,17].

It is worth mentioning that $\vec{E}$ and $\vec{H}$ waves are symbiotic. In other words, they neither can occur without the other. In this sense, $\vec{E}$ is fluctuating in time creates a magnetic flux, ϕ changing in the space and time (Fig. 1.7). In this circumstance, an

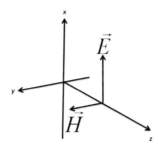

Figure 1.6. Electric and magnetic fields are propagated perpendicular to each other.

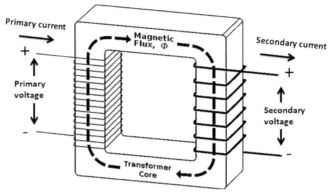

Figure 1.7. The electric vector creates magnetic flux.

electric field is produced by the changing in a magnetic field in the space and time. This reciprocated a generation of electric and magnetic field consequences in the transmission of the electromagnetic wave [18].

Subsequently $k(\omega\mu)^{-1}$ is a real number, $\vec{E}$ and $\vec{H}$ are in the phase. In other words, if $\vec{H}$ is zero, then $\vec{E}$ is zero too. Equally, when $\vec{E}$ achieves its maximum value, $\vec{H}$ also accomplishes its maximum value. In addition, Maxwell's equations elucidate the superposition impression consistent with which the resultant movement created by two liberated instabilities is the vector sum of the shift generated by the disorders autonomously. In other words, Maxwell's equations are considered linear in $\vec{E}$ and $\vec{H}$. In the circumstance, if $(\vec{E}_1, \vec{H}_1)$ and $(\vec{E}_2, \vec{H}_2)$ are two liberated solutions of Maxwell's equations, in that case, $(\vec{E}_1 + \vec{E}_2, \vec{H}_1 + \vec{H}_2)$ will also be a solution to Maxwell's equations. Thus, the superposition theory is a consequence of the linearity of Maxwell's equations. In this view, the field associated with the electromagnetic wave is extremely showed that the dielectric permittivity ε relies on $\vec{E}$. Under this circumstance, Maxwell's equations will become nonlinear. Nonetheless, the superposition theory will not remain valid anymore. In other words, the superposition theory does not hold based on the nonlinearity. Lastly, the plane wave of electromagnetic wave propagation is considered as linearly polarized because $\vec{E}$ is always along the x-axis and similarly, $\vec{H}$ is always along the y-axis (Fig. 1.8). These are well known by vertical and horizontal polarization, respectively [2,12,15,17,20].

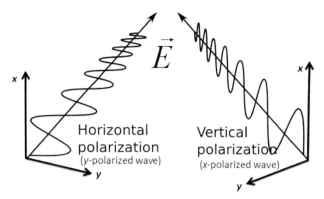

Figure 1.8. The Electric field variation of *x*-polarized wave and *y*-polarized wave.

1.5 The Poynting Theory

Let assume the case of an electromagnetic field restrained to a specified volume. How does the energy-restricted in the field change? There are two procedures by which it can occur. The first is by the mechanical work done by the electromagnetic field of the currents, which would appear as Joule heat and the second procedure is by the radiative flux of energy across the surface of the volume. In this sense, let us consider Equations 1.3 and 1.4 for linear material [17]:

$$\vec{H}.\frac{\partial \vec{B}}{\partial t} + \vec{E}.\frac{\partial \vec{D}}{\partial t} = 0.5\frac{\partial}{\partial t}(\vec{B}.\vec{H} + \vec{D}.\vec{E}) \tag{1.36}$$

In the case of

$$\nabla.(\vec{E}x\vec{H}) = -\vec{H}.\frac{\partial \vec{B}}{\partial t} - \vec{E}.\frac{\partial \vec{D}}{\partial t} - \vec{J}.\vec{E} \tag{1.37}$$

Then

$$\text{div } \vec{S} + \frac{\partial u}{\partial t} = -\vec{J}.\vec{E} \tag{1.38}$$

where

$$u = 0.5(\vec{B}.\vec{H} + \vec{D}.\vec{E}) \tag{1.38.1}$$

The quantity 0.5 $(\vec{B}.\vec{H} + \vec{D}.\vec{E})$ denotes the magnetic and electric energies per unit volume, respectively. Equation 1.38 resembles the equation of continuity. In this regard, if a charge q moving with velocity v and is acted on by $\vec{E}$ and $\vec{H}$, then the work is done by $\vec{E}$ and $\vec{H}$ across the distance $d\vec{s}$ would be $\vec{F}.d\vec{s}$. Consequently, $\vec{F}$ is the force on the charge due to $\vec{E}$ and $\vec{H}$ can be given by [1,3,17]:

$$\vec{F} = q(E + v \ x \ B). \tag{1.39}$$

Thus, the work is done per unit time can be formulated as:

$$\vec{F}.\frac{ds}{dt} = q\vec{E}.v \tag{1.40}$$

Let us assume that there are N charged particles per unit volume, each carrying a charge q. In this case, the mathematical expression of the work is done per unit volume can be written as:

$$Nqv.\vec{E} = \vec{J}.\vec{E} \tag{1.41}$$

while $\vec{J}$ denotes the current density. In this view, the energy generates in the form of the kinetic energy of the charged particles. Subsequently, the term $\vec{J}.\vec{E}$ signifies the well-known Joule loss, and thus, the quantity $\vec{J}.\vec{E}$ can denote the rate as a result of which energy is formed per unit volume per unit time. In this understanding, the mathematical description of the plane wave in a dielectric can be written as:

$$\vec{E} = \vec{x}E_0 \cos(\omega t - kz) \tag{1.42}$$

and,

$$\vec{H} = \vec{y}H_0 \cos(\omega t - kz) = \vec{y}\frac{k}{\omega\mu}E_0 \cos(\omega t - kz) \tag{1.43}$$

The term $\dfrac{k}{\omega\mu}E_0 \cos(\omega t - kz)$ denotes the amount of energy along the electromagnetic wave propagation in the z-direction.

Thus, the Poynting vector can be mathematically written as:

$$\vec{S} \equiv \vec{E}x\vec{H} \tag{1.44}$$

The Poynting vector implies to determine the amount of energy that crosses a unit area, i.e., perpendicular to the z-axis per unit time. This can mathematically described by:

$$\left\langle \vec{S} \right\rangle = \vec{z}\frac{k}{2\omega\mu}E_0^2 \tag{1.45}$$

Here, $\langle .... \rangle$ represents the time-average of the magnitude within the angular brackets. Equation 1.46 can be written as a function of dielectric permittivity ε_0 as follows:

$$\left\langle \vec{S} \right\rangle = \vec{z}0.5\varepsilon_0 cnE_0^2 \tag{1.46}$$

The Poynting vector is significant since it aligns the three vectors of an electromagnetic wave: the electric field, the magnetic field and the direction of propagation. These three vectors are reciprocally perpendicular; that is, each is perpendicular to the other two. Their relative array is resolute by the right-hand regulation of the cross product (that is; the $\times$ between $\vec{E}$ and $\vec{H}$ in Equation 1.44). In this circumstance, the intensity of the electromagnetic beam I can be given by [4,11,17,19]:

$$I = v0.5\varepsilon_0 E_0^2 = cn^{-1}\left[0.5\varepsilon_0 n^2 E_0^2\right] \tag{1.47}$$

Thus, the energy density of the electric field is the same as the energy density of the magnetic field everywhere in the electromagnetic wave. Consequently, we can express the instantaneous power per unit area of an electromagnetic wave in terms of the magnitude of the electric field or that of the magnetic field. Finally, the magnitude of $\vec{S}$ is correlated to the immediate rate at which energy is conveyed by an electromagnetic wave over a specified area, or more simply, the immediate power per unit area. Consequently, the units of the Poynting vector are thus watts per square meter (W/m²). For instance, the value of $\vec{S}$ invisible light range from a few W/m² in a faint light of a candle to many MW/m² in the most intense laser beams [5,17,20].

Needless to say, the energy transported by an electromagnetic wave is proportional to the square of the amplitude, E^2, of the wave. The Poynting vector is the energy flux vector. It is named after John Henry Poynting. Its direction is the direction of propagation of the wave, i.e., the direction in which the energy is transported. In other words, in an electromagnetic field, the flow of energy is given by the Poynting vector (Fig. 1.9). This vector is in the direction of propagation and accounts for radiation pressure, for an electromagnetic wave.

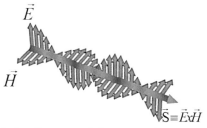

Figure 1.9. Poynting vector of electromagnetic wave propagation.

1.6 Waves from Localized Sources

When a wave originates in a localized source such as an atom, a radio transmitting antenna, a light bulb or a star, its wave fronts are not planes but expanding spheres. As it expands, the wave's energy is spread over the area of an even-larger sphere—whose area increases as the square of the distance from the source. Therefore the power per unit area—the intensity—decreases as the inverse square of the distance [1,5,17]:

$$\vec{S} = \frac{\vec{P}}{4\pi\vec{r}^2} \tag{1.48}$$

Here $\vec{S}$ and $\vec{P}$ can be either peak or mean intensity and power, respectively, and r is the distance from the source [20]. The intensity decreases not because electromagnetic waves "weaken" and lose energy, but because their energy becomes spread with time thinly. This because the intensity of an electromagnetic wave is proportional to the square of the field strengths (Equations 1.45), Equation 1.48 spectacles that the fields of a spherical wave decrease as r^{-1}. On the contrary, the r^{-1}

decreases in the electric field of a stationary point charge, and you can see why the wave fields associated with an accelerated charge dominate in all, but the immediate vicinity of the charge [5,19].

1.7 Momentum as the Route of Radiation Pressure

Dynamic objects are a function of energy and momentum. Accordingly, so are waves. Maxwell revealed that wave energy E and momentum M are allied with:

$$M = Ec^{-1} \tag{1.49}$$

The wave intensity $\bar{I}$ is the average amount at which the wave transfers energy per unit area. Therefore, the wave conveys momentum per unit area at the amount $(\bar{I}c^{-1})$. In this view, an object that absorbs the wave energy, for instance, a black body exposed to sunlight, which also absorbs this momentum [5,9,17,20]. Consistent with Newton's law, $F = \dfrac{dM}{dt}$ the black body, then practises a force F. Since $\bar{I}c^{-1}$ is the rate of momentum absorption per unit area, the result is a radiation pressure P_{rad} is given by:

$$P_{rad} = \bar{I}c^{-1} \tag{1.50}$$

In this view, P_{rad} doubles when an object imitates electromagnetic waves. In other words, P_{rad} is just like rebounding a basketball off the backboard alterating the ball's momentum by $2mv$ (Fig. 1.10). Therefore, P_{rad} delivers momentum $2mv$ to the backboard. The pressure of regular light is insignificant and challenging to determine. However, high-intensity lasers can essentially escalate tiny particles [10–20].

Needless to say, the concept in which, electromagnetic waves convey momentum acts as a tremendous role in Einstein's development of his famous formula $E = mc^2$.

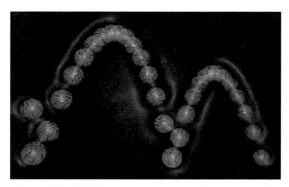

Figure 1.10. Multiple exposure sequence of a bouncing ball.

1.8 Can Maxwell's Electrodynamics Formulate in Space and Time?

It is a critical question to answer. Maxwell's electrodynamics can be formulated in terms of Minkowski space-time. Let assume different satellite data acquired over the same area with different months, for instance, from January till October. These satellite data are then stacked to provide us with a 3-D three space-time (Fig. 1.11).

In this regard, the considerations of electromagnetic waves leads to the theory of relativity in the first stage. Indeed, Maxwellian electrodynamics is considered as a basic law of relativity physics. Therefore, it defines light rays as electromagnetic waves. In this sense, the trajectory of a light ray in a vacuum can be placed on the light-cone. Thus, the speed of light becomes a diagonal line in the light-cone diagram. We can therefore present the electromagnetic waves coded in satellite data with a different time in the light-cone diagram (Fig. 1.12) [6,9,18].

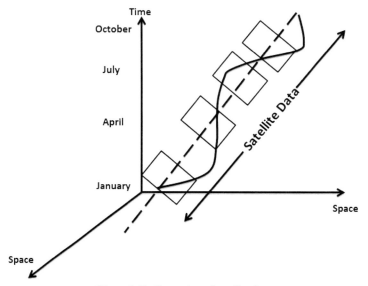

Figure 1.11. Space-time of satellite data.

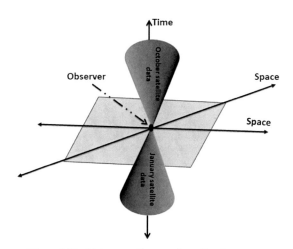

Figure 1.12. Light-cone diagram of satellite data.

In a light-cone diagram, we can determine the changes of the object in time series of satellite data. In other words, the trajectory curvature of an object changes can be explained in space and time with different satellite data time acquisitions. This kind of plot is often referred to as a graph in space-time, reflecting how these two dimensions satellite data are entangled in relativity theory. This is important for forecasting oil spill trajectory movement from the different time of the satellite data. This will be elucidated extremely through this book.

Quantization of Maxwell's Equations and Electromagnetic Field

2.1 Definitions of Quantization of Electromagnetic Field

It is well known that the electromagnetic field involves discrete energy parcels, which are called photons. In this sense, photons are massless particles of a certain energy, a certain spin, and a certain momentum. Conversely, the quantization of the electromagnetic field is defined as the quantity of the discrete energy of photons.

Let assume that K is the wave vector, which is used to describe the propagation of waves. In this view, the photons act like waves and have the propagation frequency f. Conversely, the wave vector has a magnitude and direction. The magnitude presents either the wave number or angular wave number of the wave, which is inversely proportional to the wavelength. On the other hand, the wave direction presents the direction of the wave propagation. In this case, the general equation to quantize the electromagnetic field can be mathematically written by:

$$m_{photon} = 0 \tag{2.1}$$

$$H\left|\vec{K},\mu\right\rangle = hf\left|\vec{K},\mu\right\rangle \text{ with } f = c\left|\vec{K}\right| \tag{2.2}$$

$$P_{EM}\left|\vec{K},\mu\right\rangle = \hbar\vec{K}\left|\vec{K},\mu\right\rangle \tag{2.3}$$

$$S\left|\vec{K},\mu\right\rangle = \mu\left|\vec{K},\mu\right\rangle \quad \mu = \pm 1 \tag{2.4}$$

Equations 2.1 to 2.4 demonstrate that the photon has zero rest mass and propagates with energy equals $hf = hc\left|\vec{K}\right|$ where $\vec{K}$ is the wave vector and c is the speed of light. This photon has electromagnetic momentum $\hbar\vec{K}$ and polarization, which is an eigenvalue of *the z - component* of the photon spin.

2.2 Quantum Radiation

The main physical characteristics of electromagnetic waves, including light, is of a dual nature. In other words, electromagnetic waves act like waves; and produce interference and diffraction impacts when propagating through space. On the contrary, the interaction of electromagnetic beams with atoms and molecules, causes a stream of energy corpuscles; which, is known as photons or light–quanta. Conversely, the energy of each photon relies on the frequency or wavelength of electromagnetic radiation; which is defined as:

$$E_{photon} = hv = \frac{hc}{\lambda} \tag{2.5}$$

where $h = 6.636 \times 10^{-34} \, J.s$ presents the plank's constant. However, there was a slight pecularity with the theory. To make it work accurately, Plank had to suppose that the electromagnetic energy can only be emitted or absorbed in discrete units-given by multiplying the frequency of an electromagnetic beam by a tiny number, equal to 6.636×10^{-34}. From the quantum point of view, a beam of electromagnetic energy is composed of photons traveling at the speed of light c. The number of photons crossing a unit area, therefore, will be proportional to the power of the beam.

2.3 The Photoelectric Effect

The photoelectric effect discovered by the German physicist Philipp Lenard around the time, Plank was dramatically breaking light up into quanta. In the 1880s, the photoelectric effect was a phenomenon first observed as the emission of electrons, when light shines on a specific metal (Fig. 2.1). Initial research exploited high-frequency ultraviolet spectrum and verified that the brighter this light was, the larger the amount of electricity produced.

Consistent with the conventional electromagnetic principle, the photoelectric effect can be contributed to the transmission of energy from the light to an electron. In this view, a variation in the power of light would generate fluctuations in the

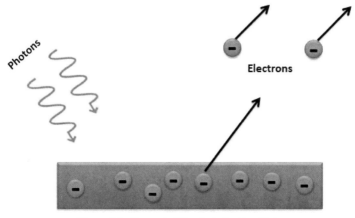

Figure 2.1. Concept of the photoelectric effect.

kinetic energy of the electrons blast out from the metal. Moreover, conforming with this concept, an adequately dim light could be assumed to present a time lag between the preliminary glowing in its light and the consequent emission of an electron. The experimental results, yet, did not relate to either of the two expectations accomplished in the conventional principle.

Initial research exploited high-frequency ultraviolet spectrum and verified that the brighter this light was, the larger the amount of electricity produced. Such relationship suited extremely with Maxwell's theory, which had recognized light as an electromagnetic wave (Chapter 1). In this regard, the more intense an electromagnetic beam was—whatever its color—the stronger the photoelectric effect should be, which was a natural magnitude of light being a wave. In this sense, when an electromagnetic beam has a lower frequency—toward the red end of the spectrum—it caused no electricity (Fig. 2.2). In this view, no matter how intense the red spectrum, it still does not cause electricity. However, a low-intensity blue spectrum can produce electricity (Fig. 2.3).

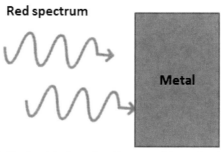

Figure 2.2. No electricity is produced because of the weak red spectrum.

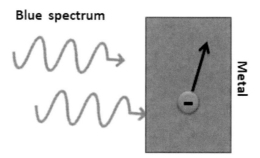

Figure 2.3. Blue spectrum produces electricity.

Planks knew his packets of light were a trick to make the calculation work. Nonetheless, Einstein dared to think that light was made of packets-particles that were later called "photons" rather than waves. In this understanding, electrons must be chunks of electricity, not irregular quantities. However, Einstein made a rather more radical interpretation of quantum radiation theory. If the single photon of an electromagnetic spectrum, instead of part of a continuous wave, has to blast an electron out its position in the metal, the photoelectric effect would function only if the photon has sufficient energy.

Research on the photoelectric effect directed to imperative stages in grasping the quantum nature of the electromagnetic beam; and electrons; and influenced the development of the theory of wave-particle duality. Other phenomena where light affects the movement of electric charges including the photoconductive effect (also known as photoconductivity or photoresistivity), the photovoltaic effect and the photoelectrochemical effect.

2.4 De Broglie's Wavelength

In 1924, Louis-Victor de Broglie articulated the de Broglie theory, declaring that all matter, not just light, has a wave-like nature; he correlated wavelength (λ), and momentum (P_{EM}): $\lambda = hp^{-1}$. He stated that since the momentum of a photon is given by $P_{EM} = Ec^{-1}$ and the wavelength (in a vacuum) by $\lambda = cv^{-1}$ where c is the speed of light in vacuum.

De Broglie's concept illustrated just how far quantum physics of an electromagnetic beam was instigated to comprehend old postulations. Electrons were components of matter—of substance—while the photons were insubstantial electromagnetic beaming. Nonetheless, they behave in some circumstances as if they are waves and in others as particles. In this understanding, this postulation, thereby, fitted well with Bohr's model of the atom. Here, an electron could occupy only specific orbits around the nucleus, as if it were running on tracks- and jumped between these orbits in a quantum leap as it gained or lost energy in the form of a photon. But what determined, which orbits were acceptable and which were not? Bohar had revealed that electrons could be in only specific orbitals, with leaps between. Orbitals could only be present if they could retain integer numbers of wavelengths (Fig. 2.4).

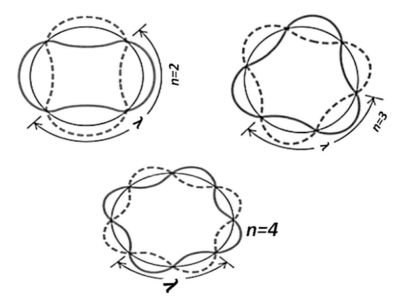

Figure 2.4. De Broglie's wavelength concept.

2.5 Quantum Electrodynamics

It is incorrect to isolate the quantum field theory from the electromagnetic wave theories. Indeed, both fields are compiled to present the field of quantum electrodynamics. In this sense, quantum electrodynamics, which refers to a quantum field theory of the electromagnetic force. The electromagnetic field is a keystone to comprehend the remote sensing technology. The main theory of remote sensing; therefore, is based on the photoelectric effect. In other words, the concept of energy in a photon is very important for remote sensing technology. Indeed, the photoelectric effect was the first example of a quantum phenomenon to be seen at the end of the 19th century [21]. In this understanding, photons played a key part in the quantum theory of the electromagnetic field, which is known as quantum electrodynamics—not only serving as the particles of electromagnetic radiation; but also as the carrier of the electromagnetic force. Any electromagnetic phenomenon; for instance, the attraction between two opposite electric charges or dual magnets repelling one another, is understood through the vital exchange of the photon particles.

2.6 Force Carriers

It was already known that electromagnetic waves could also be thought of as particles called photons. Consistent with the Dirac equation, these particles were also the quanta of the electromagnetic field, exploited to denote the tiniest boosts of the field energy. When dual electrically charged particles combine with one another, the signal of each is affected by the electric field created by the other. Also, as stated by the Dirac equation, they achieve this procedure by the mutual conversation of photons. Photons are thought to be the "force carriers" of the electromagnetic waves/fields. The Dirac equation; therefore, can mathematically be written as:

$$\left(\beta m c^2 \left(\sum_{n=1}^{3} \alpha_n p_n\right)\right)\psi(x,t) = i\hbar \frac{\partial \psi(x,t)}{\partial t} \tag{2.6}$$

With space-times coordinates x,t; the propagation of the electron mass m as a function of the speed of light c; which, is presented with the wave function $\psi(x,t)$. The speed of light; and the Plank constant $\hbar$ are reflected in special relativity; and quantum mechanics, respectively. Further, Equation 2.6 demonstrates that the electron mass m fluctuates in space-time as a function of momentum p_n in the Schrödinger equation.

α and β are known as the Dirac matrices satisfying the relations [33]:

$$\beta^2 = \alpha_i^2 = 1 \tag{2.7}$$

$$\alpha_i \alpha_j = -\alpha_j \alpha_i \tag{2.8}$$

$$\alpha_i \beta = -\beta \alpha_i \tag{2.9}$$

where α_i and β are not simple complex numbers, which they obey a Clifford algebra and must be extracted in a matrix form. In one- and two-dimensional spatial space, they are at least 2 X 2 matrices. The Pauli matrices σ_i, i.e., $i = x,y,z$ and satisfy all these relations:

$$\{\sigma_i,\sigma_j\} = 2\delta_{ij} \tag{2.10}$$

Here

$$\sigma_x = \begin{pmatrix} 0 & 1 \\ 1 & 0 \end{pmatrix}, \quad \sigma_y = \begin{pmatrix} 0 & -i \\ i & 0 \end{pmatrix}, \quad \sigma_z = \begin{pmatrix} 1 & 0 \\ 0 & -1 \end{pmatrix} \tag{2.11}$$

Therefore in one-dimension (1-D), the two Dirac matrices α_x and β are somewhat dual of the three Pauli matrices, for instance [33],

$$\alpha_x = \sigma_x, \quad \beta = \sigma_z \tag{2.12}$$

In two dimensions, therefore, the three Dirac matrices are the Pauli matrices:

$$\alpha_x = \sigma_x, \quad \alpha_y = \sigma_y, \quad \beta = \sigma_z \tag{2.13}$$

However, in 3-D, we cannot acquire more than three 2 X 2 matrices sustaining the anticommutation relationships [32–34]. Consequently, the four Dirac matrices are at least four-dimensional (4-D) and can be formulated in terms of the Pauli matrices:

$$\alpha_i = \begin{pmatrix} 0 & \sigma_i \\ \sigma_i & 0 \end{pmatrix} \equiv \sigma_x \otimes \sigma_i, \tag{2.14}$$

$$\beta = \begin{pmatrix} \sigma_0 & 0 \\ \sigma_i & -\sigma_0 \end{pmatrix} \equiv \sigma_z \otimes \sigma_0, \tag{2.15}$$

here σ_0 is a 2 x 2 identity matrix [32]. In general, the Dirac equation pronounces the presence and vanishing of photons through space (x); which, in turn, described an undulation of the electromagnetic field.

The Dirac equation did inordinate work of forecasting the performance of the electrons. Nonetheless, the Dirac equation also indicated that electrons could have both positive and negative energy levels. In this regard, infinite quantities of energy can be produced by quantum leaps of electrons into lower and lower negative energy (Fig. 2.5). In other words, the Dirac equation intimated that the universe enclosed an infinitely hidden sea of negative-energy, fulfilling entirely the probable spaces.

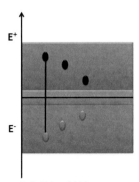

Figure 2.5. Dirac hidden sea concept.

Because of Pauli's exclusion principle, this signified an electron could never fall down into negative energy. There was an exclusion to the space chocks which are full of electrons. Sporadically, a negative-energy electron would be increased to positive energy by receiving an electromagnetic beam, leaving behind a hole (Fig. 2.6) that could be fulfilled by an ordinary electron plunging into it.

Though seeking a hole in an immeasurable sea might be a difficult task, Dirac realized that a lost negative-energy electron would seem indistinguishable to an obvious positive energy particle. Conversely, Dirac originally thought this was the proton, but soon recognized this positively charged particle required to have the identical mass as the electron. Thus, this concept was known later as antimatter-and the anti-electron or positron, should be measurable since it would have the equivalent mass as an electron. Nonetheless, it would be curved in the disparate direction by an electromagnetic field (Fig. 2.7).

In other words, the energy of a photon is converted into the mass of an electron and positron. Nonetheless, this reaction is required by the nucleus close to it to tolerate the conversion of momentum. The Dirac equation, conversely, is the simplification of the Schrodinger equation for the relativistically precise association between energy E_k and momentum $\bar{p}$, which leads to negative energy statuses and antiparticles (Fig. 2.8).

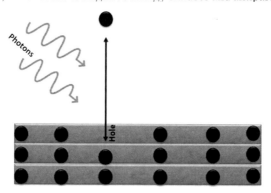

Figure 2.6. Electron hole's concept.

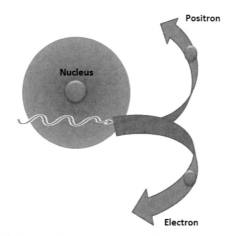

Figure 2.7. The energy of a photon is converted into a positron and electron.

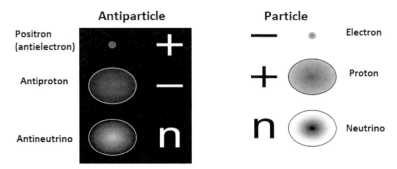

Figure 2.8. List of particle and antiparticle.

2.7 Maxwell Photon Wave Function

Dirac's particle approach; when utilized to massless spin-one particles (Fig. 2.9), i.e., single electron, precedes precisely to Maxwell's equations. In other words, the Maxwell field is considered as the quantum wave function for a particular photon. In this view, the well-known Dirac theory of quantum is accomplished in the quantization of the non-operator Maxwell field of a single photon. The scientific explanation of Maxwell's equations can mathematically be written as:

$$\frac{\partial}{\partial t}\vec{E}(\vec{r},t) = c\vec{\nabla}x\vec{H}(\vec{r},t) \tag{2.16}$$

$$\frac{\partial}{\partial t}\vec{H}(\vec{r},t) = c\vec{\nabla}x\vec{E}(\vec{r},t) \tag{2.17}$$

$$\vec{\nabla}.\vec{E}(\vec{r},t) = 0 \tag{2.18}$$

$$\vec{\nabla}.\vec{H}(\vec{r},t) = 0 \tag{2.19}$$

Both *Es* and *Hs* stand for electric; and magnetic, respectively (Chapter 1). Then Maxwell's equations can be presented in term of the wave function $\psi(\vec{r},t)$, which, is given by [23]:

$$\frac{i}{c}\frac{\partial}{\partial t}\vec{\psi} = \vec{\nabla}x\vec{\psi} \tag{2.20}$$

where

$$\vec{\psi} = \vec{E} + i\vec{H} \tag{2.20.1}$$

In fact,

$$i\frac{\partial}{\partial t}\vec{E} = \left(\vec{\nabla}x\vec{H}\right)c \tag{2.21}$$

$$i\frac{\partial}{\partial t}\vec{H} = \left(-\vec{\nabla}x\vec{E}\right)c \tag{2.22}$$

Figure 2.9. Massless spin-one particles.

Consistent with Maxwell's equations above, a photon is considered as a fundamental excitation of the quantized electromagnetic field [23]. In this view, an electron can be treated as a quasi-particle. Thus, the physical properties of an electron are massless and its spin-one nature. In this case , the Dirac theory can describe the kinematic energy E_k (Fig. 2.10) of the photon based on the momentum $\bar{p} = \left(p_x, p_y, p_z \right)$ as given by:

$$E_k = \sqrt{\left(mc^2 \right)^2 + c^2 \bar{p}.\bar{p}} \tag{2.23}$$

Equation 2.23 represents Einstein's relativistic theory. This indicates that the first derivation of Maxwell's equations is a function of relativity theory. In other words, Maxwell naively revealed a precise relativistic theory; 43 years before Einstein claimed the photon's validity! Since a photon is massless, i.e., $m = 0$, its wave function $\tilde{\psi}(p, E_k)$ should satisfy:

$$E_k \tilde{\psi}(\bar{p}, E_k) = c\sqrt{\bar{p}.\bar{p}}\,\tilde{\psi}(\bar{p}, E_k) \tag{2.24}$$

This Dirac kinetic energy equation for the electron reveals a spin-one-half particle with non-zero rest mass. Moreover, Equation 2.24 explains that the wave function $\tilde{\psi}(p, E_k)$ must have two components; i.e., momentum and kinetic energy. In this view, the propagation of electron is a function of space-time $(\vec{r}, t)$. Conversely, the electron

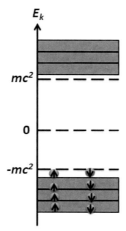

Figure 2.10. The Dirac vacuum, where all negative energy statuses are entirely dominated.

wave function is considered as a two-spinor; which is defined as $\psi(\vec{r},t) = (\psi_{1/2}, \psi_{-1/2})$; and is projected onto the quantization axis. The wave function of photon involves three components; since the photon is a spin-one particle. In this case, it is constructing three component vector fields, i.e., non-operator $\tilde{\psi}(\vec{p}, E_k) = (\psi_x, \psi_y, \psi_z)$. Let assume that $\tilde{\psi}$ is a sum of the transverse $\tilde{\psi}_T$, and e longitudinal part $\tilde{\psi}_L$. Both $\tilde{\psi}_T$ and $\tilde{\psi}_L$ obey $\tilde{\psi}_T \times \vec{p} = 0$; and $\tilde{\psi}_L \times \vec{p} = 0$, respectively. In this regard, the relevant field of a photon can be computed using the transverse part as follows:

$$E_k \tilde{\psi}_T(\vec{p}, E_k) = c i \vec{p} \times \tilde{\psi}_T(\vec{p}, E_k) \tag{2.25}$$

where $\tilde{\psi}_T(\vec{p}, E_k)$ is the amplitude function, therefore, Equation 2.25 is equivalent to Maxwell's equations. To obtain the complex value of $\tilde{\psi}_T(\vec{p}, E_k)$; then, the Fourier is used; which transforms the $\tilde{\psi}_T(\vec{p}, E_k)$ from momentum space to coordinate space, and from energy to time [23]. In this circumstance, the constraint between energy and momentum is identified by:

$$E_k = c|\vec{p}| \tag{2.26}$$

Then the momentum-space-weight function $f(E_k)$ of a photon, which is defined as:

$$f(E_k) = \sqrt{E_k} \tag{2.27}$$

This leads to present the photon energy in the coordinate-space normalization by:

$$\int d^3 r \bar{\psi}(\vec{r},t)^* . \bar{\psi}(\vec{r},t) = (2\pi\hbar)^{-3} \int d^3 p E_k(p) \tilde{\psi}^*(\vec{p}, E_k(p)).$$
$$\tilde{\psi}(\vec{p}, E_k(p)) = \langle E_k \rangle \tag{2.28}$$

where $\langle E_k \rangle$ is the expected value of the photon's energy. In this sense, the photon has no mass and has only helicity and energy, and the energy cannot strictly be localized at a point [23]. In this regard, the probability density of photon's energy is defined as the functioning ratio of:

$$P(\bar{\psi}(\vec{r},t)E_k) = \frac{\bar{\psi}(\vec{r},t)^* . \bar{\psi}(\vec{r},t)}{\langle E_k \rangle} \tag{2.29}$$

In this case, the complex Maxwell equations can mathematically be written as:

$$i\frac{\partial}{\partial t} \bar{\psi}_T(\vec{r},t) = c\bar{\nabla} \times \bar{\psi}_T(\vec{r},t). \tag{2.30}$$

Needless to say, the classical Maxwell equations are the wave equations for the quantum wave function $\bar{\psi}_T$ of a photon. In other words, the usual Maxwell field is the quantum wave function for a single photon, which it transforms as a three-dimensional vector arises from the spin-one nature of the photon. Further, the Maxwell field is considered as analogous to the Schrödinger equation wave function $\bar{\psi}(\vec{r},t)$. In this understanding, it transforms probability amplitudes for various possible quantum events; wherein, the photon energy originates within a convincing volume. In other words, the electron's location cannot be described as a realistic [23–25].

2.8 Quantanize of Electromagnetic Waves

The quantization of the electromagnetic field leads to a quantum description of electromagnetic waves propagating in a conducting linear media with time-dependent electric permittivity and conductivity. Let assume that $\bar{u}_T(\bar{r})$ presents the mode of the vector potential and $\tilde{\psi}_T(\bar{p}, E_k)$ presents the amplitude functions of each cavity mode. In this view, the integration between the Schrödinger equation and the Hamiltonian is required to quantize the electromagnetic field. To this end, the mathematical description of wave functions for each mode of the electromagnetic field is given by [26,31]:

$$\psi_n(\phi_T, t) = e^{[i\varphi_n(t)]} \left[\frac{1}{\pi^{0.5} \hbar^{0.5} n! 2^n \phi_T} \right]^{0.5} \times H_n \left[\left(\frac{1}{\hbar} \right)^{0.5} \frac{\phi_T}{\rho_T} \right]$$

$$\tag{2.31}$$

$$e^{\left[\frac{i\varepsilon_0 e^t}{2\hbar} \left(\frac{\dot{\rho}_T}{\rho_T} + \frac{ie^{-t}}{\varepsilon_0 \rho^2_T} \right) \phi_T^2 \right]}$$

Here Hn is the Hermite polynomial of order n, and ρ_T is the factor, which introduces to satisfy the normalization condition. ϕ_T; therefore, presents the canonical operator; and the phase function φ_n is defined as [26]:

$$\varphi_n(t) = -(n - 0.5) \int_0^t \frac{e^{-\tau}}{\varepsilon_0 \rho_T^2(\tau)} d\tau. \tag{2.31.1}$$

where ε_0 is permittivity; and τ time delay. Equation 2.31 demonstrates that a precise and artless quantum narrative of electromagnetic waves propagating in time-dependent conducting and nonconducting media. Furthermore, this narrative can be accomplished by combining a damped quantum-mechanical oscillator with each mode of the electromagnetic field. As a result, a fusion of the procedure was achieved to acquire the quantum behavior of electromagnetic waves in free space and cavities imparted with a (time-dependent) material medium [26]. It is frequently achieved, in the former case, by the connotation of a conventional oscillator with each mode of the quantized field. Conversely, it can be executed, in the latter one, by combining a time-dependent harmonic oscillator [26–31].

2.9 Feynman's Perspective of Electromagnetic Waves

According to the above perspective, Paul Dirac used the theory of relativity; which he delivered as a relativistic theory electron. However, the Dirac theory did not consider the effects of the electron interaction with light. In matter of fact, this theory claimed that an electron had a magnetic moment, i.e., something is similar to the force of a little magnet-which has a strength of exactly 1 in certain units. However, it was found later that the actual number is closer to 1.00118, which equals infinity by using quantum electrodynamics.

Richard Feynman believed that light comes in the form of particles. It is extremely important to know that light behaves as particles, as it behaves as waves. In this view, Feynman speculated that both photon and electron move from one position and time to another position and time. In other words, photon and electron are fluctuating in a dynamic system, which is a function of time. Moreover, an electron absorbs or emits a photon at a certain position and time. These dynamic behaviors are presented in a Feynman diagram (Fig. 2.11). Conversely, the Feynman diagram provides a wavy line for photon propagation; and a straight line for electron ejection. However, a junction of two straight lines and a wavy one for a vertex indicating emission or absorption of a photon by an electron.

In this case, another kind of shorthand for the numerical quantities is introduced by Feynman, which is well-known as a probability amplitudes. It is denoted by the square of the absolute value of total probability amplitudes (Fig. 2.12). In this regard, if a photon propagates from one place and time **A** to another place and time **B**, the probability of this movement can be written in Feynman's shorthand as $P(A$ to $B)$. The similar probability of electron moves from C to D can be identified as E (**C** to **D**). The probability can tell us about the probability amplitude for the emission or absorption of a photon; which is identified as j.

This is an approach of presenting a quantum interaction as a series of lines on the plot of space and time. In this regard, the Feynman diagram plays a tremendous role in the calculations, which usually accompany quantum physics. In fact, the diagrams are visual versions of the mathematical formulae. In this context, the probabilities of a particular outcome can be calculated as the sum of the outcomes of the different possible Feynman diagrams. In other words, these diagrams are known as Feynman propagator; which can be mathematically expressed by:

$$P(A \text{ to } B) \rightarrow D_F(x_B - x_A), \quad E(C \text{ to } D) \rightarrow S_F(x_D - x_C) \tag{2.32}$$

here a shorthand symbol, for instance, x_D presents the four real numbers that offer the position and time in 3-D of the position labeled A.

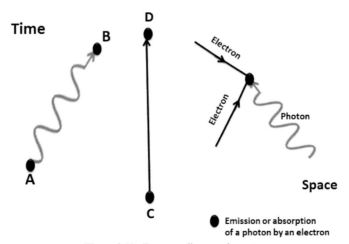

Figure 2.11. Feynman diagram elements.

$$S = \sum P$$

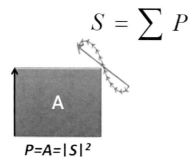

$P=A=|S|^2$

Figure 2.12. Probability amplitudes.

Feynman's work on QED showed that all the behaviors of electromagnetic beams that needed a wave description could be explained by using a particle model. The quantum particles mapped out by Feynman diagrams are not just free to take any spatial path, but can also be flexible in their route through time. For instance, a positron can be represented as an electron traveling backward in time, which is shown as the central straight line in Feynman diagram (Fig. 2.13).

A general quantum interaction is for an electron to absorb a photon of the electromagnetic wave and then produces tiny particle of single photon. This practice is accountable for each noticed from the blue sky and the reflection of light from object surface. Conversely, an electron moving backward in time revolves out to be physically identical to a positron traveling forward in time. The newly created positron—the antimatter corresponding to the electron that Dirac had expected-fuses with an original electron to create a new photon, while the electron of the pair carries on. So we begin and end with a photon and an electron. This can be a cornerstone to understand how remote sensing is operating. For instance, the antenna generates the beam of electrons moving forward in time and after interacting with objects in the ground moving backward to the antenna at a slightly different time, which is known as delay time.

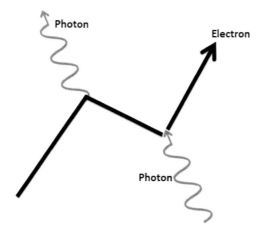

Figure 2.13. An electron traveling backward in a Feynman diagram.

Feynman was also responsible of another time-bending approach of examining quantum occurrence. Conversely, Maxwell's equations for electromagnetism had two clarifications portraying dual dissimilar sorts of an electromagnetic wave: retarded waves; which are the ones we understanding; and advanced waves, which move backward in time from the object to the antenna. The advanced wave explanation had solely been discounted since it showed to have no association with the truth. Though, Feynman commented that the advanced waves indeed exist. This, therefore, overwhelms a potential problem when an electron emanates a photon-the recoil created in the electron by the conservation of momentum. In this regard, when two photons are warped, the advanced photon propagates backward in time from the destination of a retarded photon to the electron that instigated the recoil, removing the feedback (Fig. 2.14).

In this understanding, each photon would have the energy of the absorbed one, and because of their opposing motion through time as well as space at the same time. The result cannot be distinguished from that generated by a single conventional photon. However, it makes sense of Maxwell's equations. In principle, advanced waves deliver a complex mechanism for sending a photon backward in time, nonetheless, to achieve it we would be required to locate a region of space lacking in absorbers of electromagnetic spectra. The only microwave spectrum as part of an electromagnetic beam does not absorb in the atmospheric space due to its long wavelength.

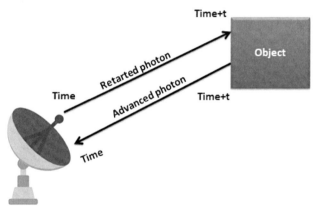

Figure 2.14. Retarded photon and advanced photon.

2.10 Feynman's Derivation of Maxwell's Equations

The physical concern is that by turning back and forth between classical notation and quantum, Feynman moves the physical postulations that are obstruct in verifying Maxwell's equations distinctively. Let us assume that a particle at the position x^μ with the momentum p^μ, which is a function of $F(x^\mu)$. Then the scientific explanation of the forces are acting on particles can mathematically be written as:

$$\frac{dp^\mu}{d\tau} = F_1^\mu(x^\mu) + F_2^{\mu\nu}(x^\mu)pv + \dots\dots \tag{2.33}$$

here F_i is tensor represents the field F. However, F_i is much too general and needs more physical assumptions to be identified. This assumption $\ell(x^\mu, \dot{x}^\mu, t)$ involves the Lagrangian; which is quadratic in velocity. The mathematical differentiation of Equation 2.33 implies that the force must be at most linear in momentum. In this view, this linear momentum is formulated as:

$$\frac{dp^\mu}{d\tau} = F_1^\mu + F_2^{\mu\nu} p\nu. \tag{2.34}$$

Equation 2.34 is determined from classical Hamiltonian mechanics as follows:

$$\left[x_i, v_i\right] = i\frac{\hbar}{m}\partial_{ij}. \tag{2.35}$$

In order to implement quantum mechanics to Equation 2.35, the Dirac prescription of replacing Poisson brackets with commutators is performed. This yields the canonical commutation relations $\left[x_i, p_j\right] = i\hbar\partial_{ij}$, where x_i and p_j are characteristically canonically conjugate. The momentum can be formulated based on Lagrangian and is determined from:

$$p \equiv \frac{\partial\ell}{\partial\dot{x}} \tag{2.36}$$

Equation 2.36 demonstrates that the Lagrangian is quadratic in velocity, subsequently, the force is at utmost linear in velocity. In the cases of $F_1 = 0$ and $F_2 = F$, Equation 2.34 can be recovered into the Lorentz force law by:

$$\frac{dp^\mu}{d\tau} = F^{\mu\nu} p\nu. \tag{2.37}$$

It is a difficult task to acquire Maxwell's equations because there is no information about the field's dynamics. However, to overcome this gap, the simplicity of the Hamiltonian is considered. So the mathematical description of the quadratic momentum is defined as:

$$\text{H} = p^2(2m)^{-1} + \bar{A}_1.\vec{p} + A_2 \tag{2.38}$$

The Hamiltonian equation of the momentum under circumstances of gathering $\bar{A}_1$ and A_2 into a four-vector A^μ is formulated as:

$$\frac{dp^\mu}{d\tau} = (dA)^{\mu\nu} p\nu \tag{2.39}$$

where d the exterior derivative. That is, the simplicity of the Hamiltonian forces the field F to be described in terms of potential, $F = dA$. Since $d^2 = 0$, so $dF = 0$, which contains two of Maxwell's equations, specifically Gauss's law for magnetism and Faraday's law.

The vital conclusion is that Feynman's derivation is countless, nonetheless not entirely known. In particular, it is not certainly collaborating classical and quantum mechanics at all—the quantum equations that Feynman exploits are corresponding to classical ones derived from the Hamiltonian equation. In fact, Feynman used the Dirac quantization approach.

It is not surprising that electromagnetism appears almost 'for free', because the space of possible theories really is quite constrained. In the more general framework of quantum field theory, we can obtain Maxwell's equations by assuming locality, parity symmetry, Lorentz invariance and that there exists a long-range force mediated by a spin 1 particle. This has consequences for classical physics, because of the only classical physics, we can observe are those quantum fields which have a sensible classical limit. In fact, nonentity has never quantified an electric field—only how it acts on particles.

2.11 Photon Spins

From the point view of quantum mechanics and particle physics, spin is defined as an inherent mode of angular momentum, which is conveyed by elementary particles, compound particles (hadrons) and atomic nuclei. Conversely, spin is one of dual sorts of angular momentum in quantum mechanics, the other being orbital angular momentum. In this regard, the orbital angular momentum is the quantum-mechanical; which is equivalent to the conventional angular momentum of orbital revolution. Further, it exists once there is a periodic configuration to its wavefunction owing to the angle variations. Let assume that the orthonormal Cartesian vectors e_x and e_y, so the spin operators can mathematically be defined as:

$$S_z \equiv i\hbar(e_x \otimes e_y - e_y \otimes e_x) \text{ and cyclically } x \to y \to z \to x. \tag{2.40}$$

The twofold operators $\otimes$ between the two orthogonal unit vectors are dyadic operator. In this context, the unit vectors are perpendicular to the propagation direction k (the direction of the z-axis, which is the spin quantization axis). Conversely, the angular momentum of the spin S_o is defined by:

$$\left[S_x, S_x \right] = i\hbar S_z \tag{2.41}$$

From the point view of dyadic operator Equation 2.41 can be mathematically expressed as:

$$\left[S_x, S_x \right] = i\hbar \left[-i\hbar \left(e_x \otimes e_y - e_y \otimes e_x \right) \right] = i\hbar S_z \tag{2.42}$$

By involving Equation 2.3, Equation 2.42 can be modified as:

$$-i\hbar \left(e_x \otimes e_y - e_y \otimes e_x \right).e^\mu = \mu\hbar e^\mu, \quad \mu = \pm 1, \tag{2.43}$$

where μ identifies the photon spin. In fact, the photon can be designated a triplet spin with spin quantum number $S = 1$. In this view, Equation 2.4 can be developed to identify the photon spin as:

$$S_z \left| \vec{K}, \mu \right\rangle = \mu\hbar \left| \vec{K}, \mu \right\rangle \quad \mu = \pm 1. \tag{2.44}$$

Equation 2.44 demonstrates that the photon has no forward ($\mu = 0$) spin component because the vector potential is a transverse field.

In specific approaches, spin is like a vector quantity; it has a well-defined magnitude, and it has also a direction. Nonetheless, quantization creates a different direction as compared to one of an ordinary vector. In other words, entirely elementary particles of a specified type have the similar magnitude of spin angular momentum, which is designated by allocating the particle a spin quantum number. In fact, a spin quantum number is a quantum number, which parameterizes the inherent angular momentum or simply spin of angular momentum of a specified particle. In other words, it designates the energy, form and coordination of orbitals. In short, each particle has its own Hilbert space, each of which satisfies the usual angular momentum commutation relations and ladder operators defined.

2.12 Do Maxwell's Equations Describe a Single Photon or an Infinite Number of Photons?

Maxwell's theory is considered as the quantum theory of a single photon and geometrical optics as the classical mechanics of this photon. For instance, the conventional constraint of the quantum theory of radiation is accomplished when the quantity of photons develops. In this sense, the occupation quantity can as well be considered as an endless fluctuating. The space-time development of the classical electromagnetic wave approximates the dynamical behavior of trillions of photons. Since the photons do not interrelate to very reliable approximation for frequencies lower than $m_e c^2/h$ (me = electron mass), the theory for one photon corresponds reasonably to the theory for an infinite number of them, modulo Bose-Einstein symmetry concerns.

In other words, the single photon behavior can be described by Maxwell's equations through involving the Fourier transform version of Maxwell's equations. The real space-time version of Maxwell's equations would require an examination of a superposition of an infinite number of photons by using an inverse Fourier transform. Further, this can be considered by using Feynman diagrams. In fact, conventional electromagnetism is described by a subset of the tree-level diagrams, while quantum field theory requires both tree level and diagrams that have closed loops in them. It is a fact that the lowest mass particle photons can create a closed loop by interacting with, the electron, that retains photons from scattering off each other.

In general, the single photon wave function is a solution of the free Maxwell equations in vacuum, and any non zero solution of the free Maxwell equations in a vacuum is a possible single-photon wave function. In other words, the mode of a coherent state of (arbitrarily many) photons is a solution of the free Maxwell equations in vacuum, and any non zero solution of the free Maxwell equations in a vacuum is a possible mode of a coherent state of photons. Conversely, the zero solution corresponds to the coherent state of zero intensity, usually called the vacuum state. More generally, for any free bosonic field theory (relativistic or not) there is a 1-1 correspondence between modes of coherent states and state vectors (wave functions) of the 1-particle Hilbert space of the theory.

Quantum Signals at Microwave Devices

3.1 Electromagnetic Wave and Microwave Beam

The previous chapters demonstrated that Michael Faraday and James Clerk Maxwell flourished in inventing speculation that coalesced electricity and magnetism into electromagnetism. In other words, the behavior of electromagnetism is described in a set of four equations, which are well-known Maxwell's equations. In this understanding, there are the many discriminations between electromagnetic radiation, electromagnetic spectrum, electromagnetic waves and electromagnetism speculations. In fact, electromagnetic waves vary in the quantity of energy they convey, called electromagnetic radiation. For instance, lower-frequency waves like radio waves radiate a smaller amount of electromagnetic radiation than the higher-frequency such as gamma rays. Electromagnetic waves and electromagnetic radiation, therefore, are regularly operated interchangeably. Conversely, the electromagnetic (EM) spectrum involves mainly wave frequencies, which is containing microwave, radio, visible light and X-rays. All EM waves, therefore, are made up of photons that propagate across the vacuum until they interrelate with the matter. However, some waves are absorbed and others are reflected. Though the sciences generally categorize EM waves into seven rudimentary types, these are entirely appearances of a similar phenomenon (Fig. 3.1).

Visible light waves allow humans to see the world around them. In this regard, the diverse frequencies of visible light are qualified by people like the colors of the rainbow. The frequencies change from the lower wavelengths, sensed as reds, up to the higher visible wavelengths, identified as violet hues. The utmost manifest natural cause of visible light is, of course, the sun. In this view, objects are observed as changing colors based on which wavelengths of light an object absorbs and which it reflects.

Ultraviolet waves have even shorter wavelengths than visible light. UV waves are the source of sunburn and can trigger cancer in living organisms. High-temperature processes release UV rays, which they can be perceived through the universe from every star in the sky. Detecting UV waves assist astronomers, for instance, in investigating the construction of galaxies.

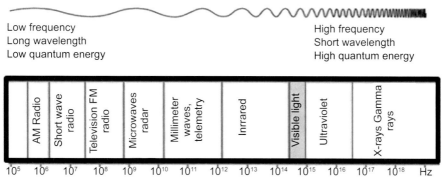

Low frequency
Long wavelength
Low quantum energy

High frequency
Short wavelength
High quantum energy

AM Radio | Short wave radio | Television FM radio | Microwaves radar | Millimeter waves, telemetry | Inrrared | Visible light | Ultraviolet | X-rays Gamma rays

10^5 10^6 10^7 10^8 10^9 10^{10} 10^{11} 10^{12} 10^{13} 10^{14} 10^{15} 10^{16} 10^{17} 10^{18} Hz

Figure 3.1 Electromagnetic spectrum.

X-rays are tremendously high-energy waves with wavelengths between 0.03 and 3 nanometers—not much longer than an atom. X-rays are emitted by sources, creating extremely high temperatures like the sun's corona, which is abundantly hotter than the surface of the sun. Natural sources of x-rays comprise extremely energetic cosmic phenomena such as pulsars, supernovae and black holes. X-rays are commonly used in imaging technology to view bone structures within the body.

Gamma waves are the highest-frequency EM waves and are released with only the most energetic cosmic objects such as pulsars, neutron stars, supernova and black holes. Terrestrial sources include lightning, nuclear explosions and radioactive decay. Gamma wave wavelengths are measured on the subatomic level and can actually pass through the empty space within an atom. Gamma rays can destroy living cells; fortunately, the Earth's atmosphere absorbs any gamma rays that reach the planet.

Radio waves are the first lowest-frequency waves in the EM spectrum. Radio waves can be exploited to convey other signals to receivers that consequently decode these signals into operational information. Countless matters, both natural and man-made, radiate radio waves. Everything that radiates heat produces radiation through the entire spectrum, but in diverse quantities. For instance, stars, planets and other cosmic bodies discharge radio waves. Radio and television stations and cell phone companies all produce radio waves that transmit signals to be obtained by the antennae in a television, radio or cell phone.

A microwave beam is electromagnetic radiation with wavelengths larger than those of visible violet light, thermal spectrum and ultraviolet light. In this connection, microwaves plunge in the variety of the electromagnetic spectrum between radio and infrared. They have frequencies from about 3 GHz up to about 30 trillion hertz, or 30 terahertz (THz), and wavelengths of about 10 mm (0.4 inches) to 100 micrometers (μm), or 0.004 inches. Microwaves are depleted for high-bandwidth communications, radar and as a source of warmth for microwave ovens and industrial applications [21]. Due to their higher frequency, microwaves can penetrate obstacles that interfere with radio waves, such as clouds, smoke and rain. Microwaves carry radar, landline phone calls and computer data transmissions as well as cook your dinner. Microwave remnants of the "Big Bang" radiate from entirely directions through space.

Practically, electromagnetic waves are created by oscillating electric charges. They vary in wavelength and frequency. Nonetheless, entirely electromagnetic waves have a similar velocity in a vacuum, approximately 300,000 km/sec. This is the speed of light, and it was the investigation and validation that light is an electromagnetic wave.

3.2 Photon of Microwave Beams

Consistent with Devoret and Schoelkopf [35] photons of microwave beam in the band 3–12 GHz (25–100 mm wavelength) have energy almost of 10^5 lesser than those of visible light. Nonetheless, at a temperature 2×10^4 lesser than room temperature, regularly present and practicable with commercial dilution refrigerator. In fact, it is conceivable to perceive and process signals whose energy are corresponding to that of single microwave photons.

The scientific explanation of microwave photons can be described by the experiment achieved by Govenius et al. [51]. In this context, a gold-palladium nanorod is coupled with superconducting aluminum islands (Fig. 3.2), to generate a series of proximity-induced Josephson junctions. Since a microwave pulse is received by the nanorod, the passing heat is increased as the pulse is absorbed and flicks the heat of the nanorod from one metastable state to another. The abrupt heat increment is transduced into a robust electrical signal at the superconducting junctions and is declaimed to identify the microwave signal. In other words, microwave photons can be detected through the heat they discharge on the absorption stage.

Microwave photons envelopes with a comparative bandwidth (Fig. 3.3) of a limited percent at carrier frequencies of a limited GHz, which can be operated with much greater relative precision than their equivalent at a few hundred of THz. In fact, microwave creators have better short-term rather than lasers, and also because microwave constituents are mechanically extremely steady, especially when cooled, compared with conventional photosensitive constituents. Furthermore, on-chip circuitry of single-photon microwave electronics can be considered in the consolidated constituent system and subsequently, the mechanism of spatial mode structure is additionaly realized than in the optical field. Conclusively, there exists an artless,

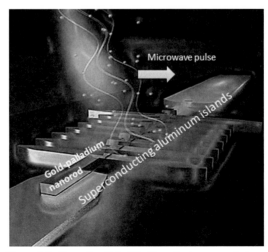

Figure 3.2. Example of identification of microwave photons.

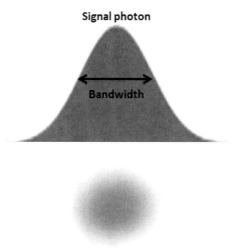

Figure 3.3. Envelopes of the signal photon.

robust, non-dissipative constituent, for instance, the Josephson tunnel junction, whose non-linearity can govern over the linear physical characteristics of the electrical circuit at the single photon intensity.

3.3 Concept of Generating Microwave Beams

Magnetism is a quantum phenomenon. Nevertheless, most magnets can be defined by conventional physics since they involve large numbers of spins. For instance, the coupling between a magnet and an electromagnetic wave can, in many circumstances, be entirely expressed by Maxwell's equations. The quantum nature of a magnet can, conversely, be divulged if it is powerfully coupled to a light field, for instance, that in a cavity. The starting point is to formulate the electromagnetic field in the cavity in a quantum state, such that the quantity of photons in the field is a worthy quantum number. "Switching on" robust coupling between the cavity and the magnet contains aligning a resonance frequency of the magnet with that of the cavity (Fig. 3.4). In these

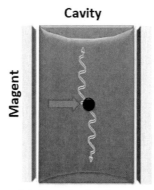

Figure 3.4. Example of a microwave cavity.

circumstances, energy in the cavity—and quantum information—can be transferred to the magnet. Once the coupling is turned off, the quantum information is "gathered" in the magnet. Switching on intense coupling over and over again relocates the situation back into the microwave cavity, where it can be detected with microwaves. The core stone requirement of this procedure is that the transmission of energy is faster than the decay mechanisms in either the microwave resonator or the magnet [52].

Bai et al. [53] stated that an intense interface between light and a magnet can manufacture a hybrid regime with energy intensities unlike from either the light or the magnet on its own. In this "robust coupling" system, quantum information can be simply shifted from the light to the magnet and vice versa. In this understanding, a strong coupling can be detected directly from a magnet, which is derived in a microwave cavity (Fig. 3.4) [53]. Conversely, a direct electrical signal can be acquired directly from the magnet instead of determining the state of the magnet circuitously with microwaves [54]. This signal is created by the professed spin pumping influence, which befalls at the crossing point between a magnet and a metal [54]. When the magnet is energized by the cavity, it diminishes towards the equilibrium by shifting a spin-polarized current into the contiguous metal layer. This spin current is conveyed by a diagonal charge current, i.e., a result of the inverse spin Hall influence and this spin pumping signal can be detected by conventional electronics [55]. Therefore, the oscillating magnetic field of the microwaves utilizes a torque on the ordered spins in the magnet and electrifying their precision. Consequently, the magnet's resonant frequency is tuned to that of the cavity by inserting the magnet structure into an aluminum microwave cavity and using an external magnetic field. As a result, the microwave spectrum is generated and measured from the cavity. Furthermore, the spectrum of the electrical spin pumping signal from the magnet (Fig. 3.5).

Both the electrical signal and the cavity have precise resonances when the magnet and the cavity are detached. Nonetheless, when the dual constituents are intensely coupled, these resonances escalate and split into dual. This phenomenon is known as an "avoided level crossing" that arises when dual regimes with matching energies

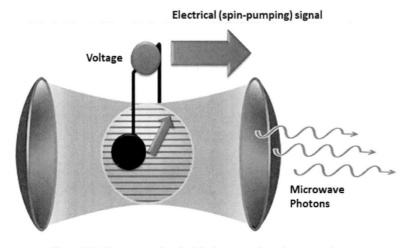

Figure 3.5. The magnet-electrical device generating microwave photons.

interrelate. Unlike the peaks in the microwave spectrum, nonetheless, the peaks in the electrical spectrum turn out to be lopsided in the intensive-coupling system. This difference, therefore, is anticipated, since the spin pumping is delicate merely to magnetic excitation. As a result, it provides information on the strong-coupling system only from the perspective of the magnetic regime. From the point of view of physics, the spin-pumping signal determines the intensity to which the magnet is energized in the intensive—coupling system [56].

3.4 Josephson Junctions for Microwave Photon Generations

As mentioned earlier, under the realistic description of electromagnetic waves, the Radio-Frequency (RF) energy at longer wavelengths is affected to a lesser degree by such obstacles. In other words, microwave signals spread in straight lines and are exaggerated slightly by the troposphere. They are not diverted or imitated by ionized areas in the upper atmospheric layers. Microwave beams do not instantly diffract throughout obstacles, for instance, hills, mountains and large human-made structures. Some attenuation, therefore, happens when microwave energy propagates through trees and frame houses.

The microwave band is well suited for wireless transmission of signals having a large bandwidth. This portion of the RF electromagnetic radiation spectrum encompasses many thousands of megahertz. Compare this with the so-called shortwave band that extends from 3 MHz to 30 MHz, and whose total available bandwidth is only 27 MHz. In communications, a large allowable bandwidth translates into high data speed. The shorter wavelengths, however, permit the use of dish antennas having manageable diameters. These antennas produce a high power gain in transmitting applications and have excellent sensitivity and directional characteristics for the reception of signals. Conversely, the declaration of these excitations must provide some quantity of quantum purity for the signals to transfer quantum information, which is the focus of interest in Josephson circuits.

The devices were used by Brian Josephson, who prophesied in 1962 that couples of superconducting electrons could "tunnel" right throughout the nonsuperconducting obstacle commencing one superconductor to another. In addition, Brian Josephson also prophesied the precise pattern of the current and voltage relations for the junction. As a result, the experimental investigation verified that he was enthusiatic in his endeavor. In 1973, Josephson was awarded the Nobel Prize in Physics for his work. In this regard, this work was a cornerstone to generate microwave beams.

To comprehend the distinctive and significant attributes of Josephson junctions, it is required to recognize the elementary theories and descriptions of superconductivity. In this view, if numerous metals are cooled and alloys to actual little temperatures, i.e., within 20°C or less of absolute zero, a phase transition follows. At this "critical temperature," the metal turns from what is known as the standard state, where it has electrical resistance, to the superconducting state, where there is basically no resistance to the flow of direct electrical current. The contemporary high-temperature superconductors, which are constructed from ceramic materials, reveal a similar performance but at warmer temperatures. In these circumstances, the electrons in the metal become matching. Conversely, due to the high critical temperature, the net interaction between two electrons is revolting. However, at the low critical temperature, though, the overall interaction between two electrons develops phase transition, a result of the electrons' interaction with the ionic lattice of the metal.

This incredibly trivial attraction permits them to plunge into an inferior energy level, causing an energy "gap." In this circumstance, electrons can propagate without being scattered by the ions of the lattice. As a result, current can flow. Thus, the electrical resistance occurs due to the ions scatter electrons. There is no electrical resistance in a superconductor, and therefore no energy loss. This is an essential requirement to generate microwave beams across the antenna. There is, nonetheless, a maximum supercurrent that can flow, called the critical current. Above this critical current, the material is normal. There is one other very important property: when a metal turns into the superconducting condition, it ejects entirely magnetic fields, as long as the magnetic fields are not too large.

At a Josephson junction, the nonsuperconducting obstacle, i.e., insulator, is splitting the dual superconductors, which must be exceptionally thin (Fig. 3.6). If the obstacle is an insulator, it has to be on the order of 30 angstroms thick or less. If the obstacle is nonsuperconducting, it can be as much as numerous microns thick. Until a critical current is grasped, a supercurrent can flow through the obstacle. In this view, electron pairs can tunnel through the obstacle deprived of any resistance. However, when the critical current is surpassed, another voltage can develop through the junction. That voltage will rely on time—that is, it is an AC voltage. This, in turn, instigates a lowering of the junction's critical current, producing even more regular current to flow—and a larger AC voltage. In these circumstances, the frequency of this AC voltage is approximately 500 gigahertz (GHz) per millivolt through the junction. Subsequently, if the current across the junction is less than the critical current, the voltage is zero. Once the current surpasses the critical current, the voltage is not zero but oscillates in time. Sensing and determining the modification from one state to the other is at the cornerstone of the many radar applications for Josephson junctions.

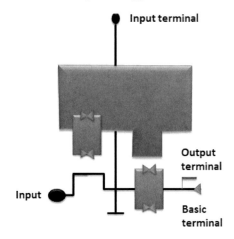

Figure 3.6. Simple concept of Josephson junction.

Electronic circuits can be made from Josephson junctions, particularly digital logic circuitry. For instance, numerous scientists are succeeding in constructing ultrafast computers utilizing Josephson logic. Josephson junctions can also be produced into circuits called SQUIDs—an acronym for Superconducting Quantum Interference Device (Fig. 3.7).

A SQUID involves a loop with two Josephson junctions intruding the loop. A SQUID is extremely sensitive to the total amount of magnetic field that penetrates the area of the loop. In this sense, the voltage is measured through the device is exceptionally powerfully interconnected to the overall magnetic field around the loop [36].

Arrays of Josephson junctions (JJA) are utilized to enhance the sensitivity of detectors, increase the output power and reduce the line width of Josephson oscillators (Fig. 3.8) [37]. This is attained not only by a sense of the intensification in impedance of the sequences array, but principally due to synchronization of Josephson oscillations in the junctions.

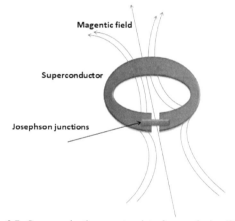

Figure 3.7. Superconducting quantum interference device (SQUID).

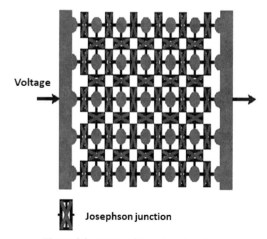

Figure 3.8. Arrays of Josephson junctions.

3.5 Mathematical Description of Quantum Microwave

Let us assume an infinite ideal transmission line which is indexed by the real number $x \in (-\infty, +\infty)$. Consider a one-dimensional electromagnetic medium, which is categorized by a propagation velocity U and a characteristic impedance Z_c of frequency ω. Therefore, the propagating wave amplitude Ψ [Watt]$^{0.5}$ is generated along the line in the $+x$ direction due to the combination of the voltage $V(x,t)$ across the line with the current $I(x,t)$ (Fig. 3.9), which is obtained from [38]:

$$\Psi^{\rightleftarrows}(x,t) = 0.5 \left[^{-1}\sqrt{Z_c}V(x,t) \pm \sqrt{Z_c}I(x,t) \right] \tag{3.1}$$

The direction of propagation is $\rightarrow$ or $\leftarrow$. Further, Equation 3.1 can be expressed in the form of the Poynting vector for a one-dimensional medium by:

$$P(x,t) = \left| \psi^{\rightarrow}(x,t) \right|^2 - \left| \psi^{\leftarrow}(x,t) \right|^2 = V(x,t)I(x,t) \tag{3.2}$$

In both Equations 3.1 and 3.2, the voltage and current follow the equivalent of Maxwell's equations as:

$$\psi^{\rightleftarrows}(x,t) = \psi^{\rightleftarrows}\left(0,t \mp \frac{x}{U}\right) = \psi^{\rightleftarrows}(x \mp Ut, 0) \tag{3.3}$$

Equation 3.3 demonstrates that the wave amplitude has the space-time translation invariance properties of traveling electromagnetic waves. The scientific explanation of the electromagnetic field of the line quantum-mechanically can mathematically be written as:

$$\psi^{\rightleftarrows}(x,t) \rightarrow \psi^{\rightleftarrows}(x) \tag{3.4}$$

The wave amplitude operator ψ is also described by Hermitian. In this regard, implementation of Fourier space can be used to describe the microwave photon as [38–40]:

$$\psi^{\rightleftarrows}[\omega] = \frac{1}{\sqrt{2\pi}} \int\limits_{-\infty}^{+\infty} dt \, e^{i\omega t} \psi^{\rightleftarrows}(t) \tag{3.5}$$

It is worth mentioning that Equation 3.5 is described ψ as the frequency domain. In other words, the time index is corresponding to a space index and therefore, the reciprocal space is corresponding to the wave vector space.

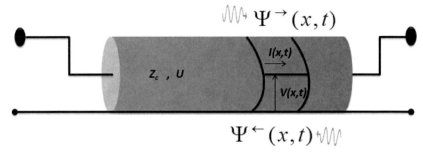

Figure 3.9. Electromagnetic transmission line implemented as a coaxial cable.

3.6 Microwave Signal Harmonic Oscillators Using Ladder Operator

Precisely, a loader operator is delineated as an operator which, when utilized in a state, generates a new state with a higher or lowered eigenvalue. Their utility in quantum mechanics results from their capability to designate the energy spectrum and associated wave functions with a more practical approach, without resolving differential equations. Conversely, the mathematics of ladder operators can easily be extended to more complicated problems, including angular momentum and numerous-particle problems. In the latter case, the operators serve as creation and annihilation operators; adding or subtracting to the number of particles in a given state. Consequently, the field ladder operators $\ell^{\pm 1}$ can mathematically be given as:

$$\ell^{\pm 1}[\omega] = \psi^{\pm}[\omega]\left[\sqrt{\frac{\hbar|\omega|}{2}}\right]^{-1} \tag{3.6}$$

In this context, ladder operators in quantum mechanics describe the incrementing (or decrementing) of a quantum number, thus mapping one quantum state into another (Fig. 3.10). This is the reason that they are often known as raising and lowering operators, i.e., ±1. Furthermore, the ladder operator concept reveals the quantum mechanical treatment of the harmonic oscillator. In other words, $\ell^{\pm 1}$ has commutation relationships, assuming a patent similarity to the ladder operators of a set of standing wave harmonic oscillators. In this view, the mathematical model of harmonic oscillators using the ladder operators is formulated as [40,42]:

$$\left[\ell^{\pm 1}[\omega_1], \ell^{\pm 1}[\omega_2]\right] = \text{sgn}(\omega_1 - \omega_2)\hat{\partial}(\omega_1 + \omega_2)\partial_{\ell^{\pm 1}} \tag{3.7}$$

Equation 3.7 does not describe the traveling of microwave photons. In doing so, the *hermitian* operator must be considered as the inverse of the Fourier transform of the ladder operators which is formulated as $\ell^{\pm 1}(t)$. In this view, $\ell^{\pm 1}(t)$ it would be appealing to deduce them as photon flux amplitude. An orthonormal signal basis of "first-quantization" wavelets [47] $w_{mp}^{\ell}(t)$ is used to identify the photons of the line as:

$$\sum_{m=-\infty}^{+\infty}\sum_{p=-\infty}^{+\infty} w_{mp}^{\ell}(t_1)w_{-mp}^{\ell}(t_2) = \hat{\partial}(t_1 - t_2) \tag{3.8}$$

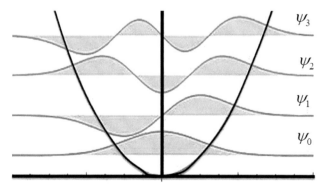

Figure 3.10. Harmonic oscillators using the ladder operators.

3.7 Quantum Electromagnetic Signals Propagating Along Transmission Lines

Here $\left(|m|, p\right) \in \mathbb{N}^+ \times \mathbb{Z}$ delineates a transmitting sequential mode of the line, and the fused amplitudes of the two equivalent wavelets, which can be realized as a fundamental degree of freedom of the microwave field. There are two conjugate wavelets per mode since the phase space of each mode is bi-dimensional. Therefore, the discreteness of the signal constituent indices is the explanation for the expression "first-quantization" and no quantum mechanics is convoluted here as entire functions are at this phase complex number evaluated. Conversely, second-quantization arbitrates when the discrete ladder field operators are defined with indices $m > 0$ and p. In this view, the photon-number operator is expressed as:

$$n_\mu = \psi_\mu^\dagger \psi_\mu \tag{3.9}$$

where $\mu = \left(\ell, |m|, p\right)$ is presented as the index of the spatiotemporal mode, and is also known as the flying oscillator. Equation 3.9 delivers the following significant marks: the complexity of the first quantization $w_{mp}^\ell(t)$ induces non-Hermitian of the photon amplitude operator ψ_μ. In contrast, the second quantization component of $\ell^{\pm 1}(t)$ is a Hermitian operator. Conversely, the frequency of a photon is vague, in contrast with what could be concluded from basic outlines to quantum mechanics. This attribute occurs once the length of the wavelet equivalent to that specific photon is not precisely associated with the inverse of the wavelet center frequency. Accordingly, the speculation of photons for a transmitting signal has to be evidently discriminated from an energy quantum. A transmitting photon is a fundamental excitation of the field conveying a quantum of exploiting and relates to a field wave function orthogonal to the vacuum. In this regard, a wavelet transform can mathematically contain several photons in modes μ_N as:

$$\left|\Psi_{n_1,\mu_1;n_2,\mu_2;\ldots;n_N,\mu_N}\right\rangle = \frac{1}{\sqrt{n_1!}}\left(\Psi_{\mu_1}^\dagger\right)^{n_1}\frac{1}{\sqrt{n_2!}}\left(\Psi_{\mu_2}^\dagger\right)^{n_2}\ldots\left|vac\right\rangle \tag{3.10}$$

where N is numerous of sequence modes $s = (n_1\mu_1; n_2\mu_2; \ldots; n_N\mu_N)$ of photon occupancy configuration, and *vac* stands for "volts of alternating current". In this view, the Alternating Current (AC) is an electric current which periodically reverses direction. In contrast to Direct Current (DC), which flows only in one direction. The abbreviations AC and DC are often used to mean simply alternating and direct, as when they modify current or voltage. It can be said that the most general wave function of the field of the transmission line(s) is a superposition of all field photon configurations in all the spatiotemporal modes of the lines [38,43]:

$$\left|\Psi\right\rangle = \sum_s C_s \left|\Psi_s\right\rangle \tag{3.11}$$

Here C_s is a quantum coefficient, which demonstrates that a state with a precisely delineated number of photons in a definite wavelet foundation can be copiously entrapped in another base. In other words, a well-defined number of photons can be explained as a coherence state, which is expressed by wavelet transform [46] as:

$$|\Psi_\sigma\rangle = e^{-\sum_\mu \left(\frac{|\sigma_\mu|^2}{2 - \sigma_\mu \psi_\mu^\dagger} \right)} |vac\rangle \qquad (3.12)$$

Here σ_μ is the set of complex coefficients, which indicates that a coherent state remains true in every wavelet basis (as can be supposed from the quadratic formula in the exponent of Equation 3.12. In this regard, a density matrix is used to describe the coherence state or coherence field. In fact, ultimate quantum field depiction means direct to the imperative notion of information enclosed in the microwave signal. As a rule, the system can be defined by a finite-dimensional Hilbert space, the Shannon–Von Neumann entropy in quantum mechanics as [42,44]:

$$S = -tr\rho \ln \rho \qquad (3.13)$$

where tr denotes the trace and ln denotes the (natural matrix algorithm. If ρ is written in terms of its eigenvectors $|1\rangle, |2\rangle, |3\rangle, \ldots$ as

$$\rho = \sum_j \eta_j |j\rangle\langle j|, \qquad (3.14)$$

then the von Neumann entropy is merely,

$$S = -\sum_j \eta_j \ln \eta_j \qquad (3.15)$$

In this form, S can be seen to amount to the information theoretic Shannon entropy. For a quantum-mechanical system described by a density matrix ρ. In this understanding, the density matrix formalism was established to expand the contrivances of conventional statistical mechanics to the quantum domain. In the conventional context, the partition function of the system can be computed in order to evaluate all possible electrodynamic quantities. Von Neumann introduced the density matrix in the context of states and operators in a Hilbert space. The knowledge of the statistical density matrix operator would allow us to compute all average quantities in a conceptually similar, but mathematically different way. Let us assume a set of wave functions $|\Psi\rangle$ that can be contingent parametrically on a set of quantum numbers $n_1, n_2, \ldots, n_N$. The natural variability, which we have is the amplitude with, which a particular wave function of the basic set participates in the actual wave function of the system. Let us denote the square of this amplitude by $\psi(n_1, n_2, \ldots, n_N)$. The aim is to turn this quantity ψ into the conventional density function in phase space. We have to prove that ψ drives over into the density function in the conventional constraint, and that it has ergodic properties. Subsequently, an examination that $\psi(n_1, n_2, \ldots, n_N)$ is an unceasing density function of the signal [40–43], an ergodic hypothesis for the probabilities $P(n_1, n_2, \ldots, n_N)$ creates ψ a function of the individual energy. In this view, one finally obtains the density matrix formalism when seeking a form where $\psi(n_1, n_2, \ldots, n_N)$ is invariant with respect to the representation used. In the form it is written, it will only yield the correct expectation values for quantities, which are diagonal with respect to the quantum numbers $n_1, n_2, \ldots, n_N$. Then the information carried by quantum signals is given by [38,44]:

$$I = \sum_{m=1}^{M} \sum_{p=-p}^{p} \log_2 \left[\sqrt{1-x^2} \left(\frac{1+|x|}{1-|x|} \right)^{\frac{|x|}{2}} \right] \left(\left\langle n_{|m|p} \right\rangle - 0.5 \right) \tag{3.16}$$

Equation 3.16 demonstrates the maximum number of excitations is unity in the domain $(|m|, p) \in \{1, 2, ..., M\} \otimes \{-P, ..., +P\}$ and that other modes are in the vacuum state. In this circumstance, a state of line characterized by an average photon number $n_{|m|p}$ per mode. The extension of these ideas to a transmission line on which a signal propagates is not trivial since the number of temporal modes is infinite and each temporal mode has a Hilbert space with the infinite dimensionality [38,45].

Consistent with the above perspective, microwave photons can be comprehended as the carriers of the information transmitted from the propagation field. In this view, microwave photons are the quanta of the electric field. Hence, electric field lines propagate at the speed of light as well as their carriers do. In this regard, photons with electric charge excrete across a field. In fact, charged particles radiate a great deal and in such random directions that it basically acts as a continuous particle flows across a field. In other words, individual photons must propagate any kind of expanse to spread other particles, so if the source of photons moves, objects affected by those photons will not be influenced by the new site until those photons are derived from a different location. As a result, photons which reveal the variation in the position "travel," and many of them can be imagined as a field.

Conversely, this essentially signifies that there are dual components of the field; the parts of the field, which reveal the original site have not yet been reorganized, and the components of the field which reveal the different site. The quantity of the field, which reveals the different site propagates out (at the speed of light), restoring the field, which was instigated by the first location. In this understanding, photons travel at the speed of light, as a result, revisions in the field occur at that speed. However, the theory of the electromagnetic field is not "reinstated by virtual photons". The electromagnetic field, therefore, basically develops as a quantum field similar to the constituents of the theory, and photons are the quanta of the field. Virtual particles, however, are not actual particles, whatever, but merely lines in a Feynman diagram.

3.8 Quantum Langevin Equation

From the point of view of physics, Langevin equation is named after Paul Langevin, which is a stochastic differential equation demonstrating the time growth of a subset of the degree of freedom. These quantities of freedom usually are combined variables differing only gradually in contrast to the other variables of the system. The fast microwave photon particles are responsible for the stochastic nature of the Langevin equation.

A decisive constituent, in the explanation of the mapping of the input microwave signal into the output signal, is the combination between the microwave circuit, which houses standing microwave electromagnetic modes and the transmission lines, which support propagating modes. This pairing is allocated hypothetically through the Quantum Langevin Equation. In this regard, consider the simplest electromagnetic circuit, which has one electromagnetic mode and is coupled with only one semi-infinite transmission line. Subsequently, the microwave signal processing occurs as a transformation of the incoming wave into the reflected outgoing wave across one-mode and one-port configuration. In this circumstance, a no-reciprocal line device is called a circulator (Fig. 3.11) must be adjoined to split the incoming and outgoing waves into dual autonomous transmission lines. In addition, this circuit configuration also delivers an approach to compute directional, across amplifiers.

Let us assume that the quantum amplitude $\vec{\psi}$ of the microwave circuit, satisfying $\left[\vec{\psi}, \vec{\psi}^{\dagger}\right] = 1$. The mathematical explanation of the incoming quantum field amplitude of the line $\vec{\psi}^{in}(t)$ at the unique port is given by:

$$\frac{d\vec{\psi}}{dt} = \frac{i}{\hbar}\left[H, \vec{\psi}\right] - 0.5k\vec{\psi} + \sqrt{k}\vec{\psi}^{in}(t) \tag{3.17}$$

Equation 3.17 is a distinctive circumstance of the Quantum Langevin Equation (QLE), where $\vec{\psi}^{in}(t)$ is the traveling photon amplitude, and H is the Hamiltonian of the circuit, written as a function of the conjugate ladder operators $\vec{\psi}$ and $\vec{\psi}^{\dagger}$. In Equation 3.17, the first term relates to that of the usual Heisenberg equation of motion for an operator in quantum mechanics. In this sense, the circuit can be arbitrarily non-linear and thus **H** will be in a broad-spectrum more complex than the simple quadratic term $\hbar\omega\vec{\psi}^{\dagger}\vec{\psi}$ of the quantum harmonic oscillator.

It is worth noting that the coefficient $\sqrt{k}$ in front of the propagating electromagnetic field amplitude represents single-handedly the fluctuation-dissipation theorem. It emphasizes the rate at which energy is radiated away from the circuit (the coefficient k of the second term) has to be compactly interconnected to the combined quantity with which random energy—radiated from the black body that the line plays the role of—distort the clarity of the state of the circuit. If one spectacle, why k seems under

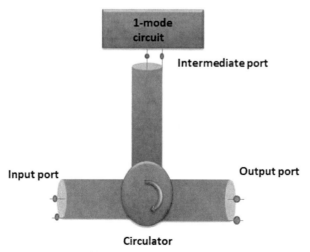

Figure 3.11. An ultimate circulator splits the input from the output.

a square root in this coupling coefficient, one just needs to remember that while is a dimensionless standing photon number amplitude, $\vec{\psi}^{in}(t)$ is the dimensioned amplitude corresponding to a photon flux (Fig. 3.12).

Subsequently, the second term in Equation 3.17 presents a damping term definitions to both the open description of the physical system, and the linear combination between the microwave circuit and the transmission line.

It is a noteworthy system involving two postulations: (i) The supposed Markov approximation, which deliberates that the combination of the system with the environment is "ohmic": for instance, the density of modes of the environment can be deliberated white through the set of circuit transition frequencies ω, as in an ideal resistance. (ii) The combination is also recognized to be feeble in the sense that k is considerably smaller than any transition frequency between the energy levels of the lumped circuit. Therefore, this is the supposed Rotating Wave Approximation (RWA) (Fig. 3.12).

Finally, the third and last term on the right-hand side of the Quantum Langevin Equation designates the function of the received microwave signal as a drive for the circuit. It denotes the equivalent of the energy loss which is modeled by the second term. Though approximations are simulated, Equation 3.17 reveres the significant commutation relationship of the ladder operators:

$$\left[\vec{\psi}(t),\vec{\psi}(t)^{\dagger}\right]=1 \tag{3.18}$$

In general, the received microwave photons have three components, which are considered on equivalent steadiness by the QLE: (i) the deterministic microwave signal to be treated, (ii) thermal or parasitic noise conveying the information-carrying signal, and (iii) quantum noise, which is the zero-point oscillations of the electromagnetic field of the semi-infinite transmission line. The inclusion of this last constituent is instigated covertly in that the Quantum Langevin Equation is an operator equation. In contrast with the Classical Langevin Equation which is just a different equation for a quantum-number function, albeit stochastic.

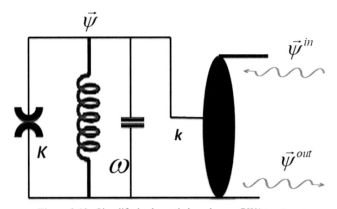

Figure 3.12. Simplified schematic based on an RWA treatment.

In the circumstance of the input signal is only a black-body noise with temperature T, the semi-infinite line is in thermal equilibrium and the mathematical description is involved in the anti-commutator grasps:

$$\left\langle \left\{ \bar{\psi}^{in}[\omega_1], \bar{\psi}^{in}[\omega_2] \right\} \right\rangle_T = \coth \frac{\hbar(|\omega_1 - \omega_2|)}{4k_B T} \hat{c}(\omega_1 + \omega_2) \tag{3.19}$$

Here, $\langle ..... \rangle_T$ is the average in the thermal state and k_B is Boltzmann's constant. In the circumstance of operating temperature is about 20 mK, Equation 3.19 spectacles that the quantum fluctuations turn into entirely prevailing over thermal oscillations at frequencies above a GHz. As it is known, $H = \hbar \omega \bar{\psi}^\dagger \bar{\psi}$, the elimination of $\bar{\psi}$ between input and output signals can be transmitted out entirely at the logical phase and one mathematically achieves:

$$\left(\frac{d}{dt} + i\omega + 0.5k \right) \bar{\psi}^{out}(t) = -\left(\frac{d}{dt} + i\omega - 0.5k \right) \bar{\psi}^{in}(t) \tag{3.20}$$

The reflection coefficient $r(\omega)$ of microwave photons can be obtained using the Fourier domain:

$$\bar{\psi}^{out}[\omega] = r(\omega)\bar{\psi}^{in}[\omega] \tag{3.21}$$

$$r_{RWA}(\omega) = -\frac{\omega - \omega_0 - 0.5ik}{\omega - \omega_0 + 0.5ik} \tag{3.22}$$

where $r_{RWA}(\omega)$ is analytic complex function characteristic of the circuit. In fact, the causation physical characteristic of the microwave circuit, which articulates the fact that it cannot create a reaction previously sent to a stimulus. In other words, its pole is in the lower half complex plane, while its zero is in the upper half angular frequency ω. In contrast, the reflection coefficient has only one pole instead of a pair is an artifact of RWA. In the circumstance linearity of the circuit, the accurate reflection coefficient can be computed exploiting a more elaborate formula of QLE deprived of RWA, while retaining the Markov approximation. Furthermore, maintaining to work in the framework of both RWA and the Markov approximation, microwave antenna can involve more than one circuit mode and more than one semi-infinite line.

As it is known, the multi-mode and multi-port generalized QLE, are mathematically described as:

$$\frac{d}{dt}\bar{\psi}_M = \frac{i}{\hbar}[H, \bar{\psi}_M] + \sum_P \left[-\frac{k_{MP}}{2}\bar{\psi}_M + \varepsilon_{MP}\sqrt{k_{MP}}\bar{\psi}_M(t) \right] \tag{3.22}$$

where M is the standing mode index, P is the port index and ε_{MP} is the rectangular matrix, which has the complex coefficients equal 1, i.e., $|\varepsilon_{MP}| = 1$. This matrix, therefore, can be determined from the particulars of the coupling of the lines to particular components of the circuit which involve capacitances or inductances, series or parallel connections.

It is worth noting that the Quantum Langevin Equation and the Input-Output Equation which together permit, at least correctly, to calculate how a given antenna circuit handles microwave signals transmitting on transmission lines, which are based on a harmonic oscillator circuit.

In this sense, the Hamiltonian of the Josephson element of superconducting tunnel junction for microwave oscillator is formulated as:

$$H = \left(\frac{\hbar}{2e}\right)^2 / L_j \cos\varphi + \frac{Q^2}{2(C_J + C_{ext})} - \frac{\hbar}{2e}\varphi.I + H_{env} \qquad (3.23)$$

The Josephson energy is presented by term of $\left(\dfrac{\hbar}{2e}\right)^2 / L_j$, the total capacitance in parallel with the Josephson element is denoted by $2(C_J + C_{ext})$, and φ the gauge-invariant phase difference across the junction. Moreover, **Q** is the charge conjugate to the phase [φ,**Q**]=2*ei*, and H_{env} is the Hamiltonian of the transmission line, which is counting the pump received signal through this channel and *I* is the current operator corresponding to the quantities of freedom of the transmission line. In this view, the amplifier functions $\langle\varphi\rangle$ with having deviations considerably less than $\dfrac{\pi}{2}$ and the cosine function in the Hamiltonian can be expanded to 4th order only, with the φ^4 term, which is considered as a perturbation [48,49]. Thus, the ladder operators of the single mode of the circuit are formulated as:

$$\varphi = \varphi_{ZPE}(\vec{\psi} + \vec{\psi}^\dagger) \qquad (3.24)$$

$$\text{Where } \varphi_{ZPE} = \sqrt{\frac{2e^2}{\hbar}} \sqrt[4]{\frac{L_j}{2(C_J + C_{ext})}} \qquad (3.24.1)$$

Equation 3.24 can be mathematically formulated in the form of the Quantum Langevin Equation as:

$$\frac{d}{dt}\vec{\psi} = -i\left(\overline{\omega} + k\vec{\psi}^\dagger\vec{\psi}\right)\vec{\psi} - \frac{k}{2}\vec{\psi} + \sum k\vec{\psi}^{in}(t) \qquad (3.25)$$

Equation 3.25 can be formulated into Hamiltonian for the degenerate parametric amplifier arising from the pumping of the Josephson junction as:

$$\frac{H}{\hbar} = \left(\overline{\omega}_a + 2k|\psi|^2\right)\partial\vec{\psi}^\dagger\partial\vec{\psi} + \left[h.c. + \frac{\mu_r\omega_0}{4}e^{i(\Omega_{aa}t+\theta)}\left(\partial\vec{\psi}\right)^2\right] \qquad (3.26)$$

where α is defined as:

$$\psi = \frac{i\sqrt{k}\psi^{in}}{(\overline{\omega}_a) + \dfrac{ik_c}{2} - k|\psi|^2} \qquad (3.26.1)$$

where c is a complex number, ψ is the classical semi signal amplitude, μ_r relative amplitude and ω_0 is the resonant frequency. Moreover, Equation 3.26 demonstrates that the center frequency of the amplifier band shifts as the growth in the pump amplitude. In this understanding, the pump tone requires to be at the center of the band for optimal amplification. Furthermore, $\Omega_{aa} = \Omega_1 + \Omega_2$ is the coupling of two pump frequencies, which facilities the use of the amplifier parameter. It is worth noting that for this device, the pump tone and the signal tone must be received by the circuit on the same port, which is difficult to deliver the widely diverse amplitude intensities of these dual waves. However, the synthetic aperture radar antenna can send the signal and receive the back scattered signal, which will be discussed later in the coming chapters.

Quantum Mechanical of Scattering Cross-Section Theory

4.1 Definitions of Scattering

Scattering is a general physical process where some forms of radiation, such as light, sound or moving particles, are forced to deviate from a straight trajectory by one or more path due to localized non-uniformities in the medium through which they pass. In this regard, the quantum mechanics describe the scattering as a function of atom-photon interaction. In this understanding, the definite detected targets are two incident photons with a certain energy, and two deviate photons with different energies. In this context, the probability of such a process (given the incident and the deviation) to occur is determined by the amplitude associated with this particular Feynman graph (Fig. 4.1). It is also worth noting that such a probability is not completely accurate

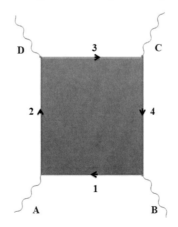

Figure 4.1. Feynman graph for scattering.

since a single diagram is just one term of the infinite formal perturbative expansion providing the transition amplitude. However, it must not interpret the Feynman graph as a collision process. Figure 4.1 reveals that Positron 1 absorbs B and becomes positron 4, which emits C and becomes positron 3. Furthermore, electron 2 and positron 3 annihilate and producing D [52–55].

However, vertices 1 could be transferred into either direction, so an equally accurate description would be:

i) B splits into electron 1 and positron 4;
ii) Electron 1 absorbs A and becomes electron 2;
iii) Electron 2 emits D and becomes electron 3; and
iv) Electron 3 and positron 4 annihilate, producing C.

Generally, the interaction of a photon and a free electron is an instantaneous process of **scattering** (transfer of momentum) between the said particles. Therefore, an interaction of a photon and an electron bound in an atom is a very fast, but is not instantaneous. In other words, the electron cloud has to restructure itself via resonant oscillation, which is the process of **absorption** in which the photon is annihilated and atom ends up in an excited state. Later, another photon can be emitted from the atom taking away the excitation. This is *not* scattering as scattering occurs for free electrons and absorption for bound.

From the point view of quantum mechanics, the photon scattering involves:

i) Elastic scattering: the kinetic energy of the scattered particles is preserved; and
ii) Inelastic scattering: kinetic energy is not conserved, e.g., due to a photon taking off with part of the energy.

4.2 Mathematical Description of Scattering Cross-Section

In this chapter, an elastic scattering is considered as the dual similar particles before and after. Since the potential velocity of a photon $U(r_1 - r_2)$ can be reduced to an effective one-body problem, where only the relative motion of the particles matter, the scattering problem can be considered in the corresponding Center-of-Mass (CM) frame. Let us assume that a uniform flux of particles with density j_{in} is incident on a scattering center S (Fig. 4.2).

In this view, the scattered photon particles are counted into a detector solid angle $d\Omega = \sin\theta d\theta d\phi$, which is enclosing the direction (θ, ϕ), respectively. The mathematical description of the scattering cross-section per solid angle is given by:

$$\frac{d\sigma}{d\Omega} = \frac{\gamma_S}{d\Omega . j_{in}} \tag{4.1}$$

Equation 4.1 reveals that the ratio amount of photons scattered γ_S into a solid angle per unit time to flux of photons with density j_{in} into solid angle $d\Omega$. In circumstance of $j_{in} = |j_{in}|$ and $\frac{d\sigma}{d\Omega}$ has a dimension area, the scattering cross-section is formulated as:

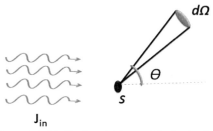

Figure 4.2. Idealized model for particle density flux.

$$\sigma = \int \frac{d\sigma}{d\Omega} d\Omega = \int_{\phi=0}^{2\pi} \int_{\theta=0}^{\pi} \frac{d\sigma}{d\Omega} \sin\theta \, d\theta \, d\phi \tag{4.2}$$

The dimension area of σ which corresponds to the entire area the incident photons are traveling through, which causes scattering. The number of photons exceeding across σ will be similarly enormous as the number of photons that eventually are scattered in a specific direction and can collect into a detector solid angle. Let us assume a hard sphere with radius R, then its area $\sigma = \pi R^2$, which is equivalent to photons scattering on a sphere of radius R. Owing to symmetry, the central potential of photon velocities $U(r) = U(|r|)$ do not rely on the azimuthal angle ϕ [60–63].

4.3 Why Does Scattering Rely on Spin?

The probability of scattering relies upon the quantity of existing quantum circumstances for the electron to scatter into, and this relies strongly on the relative track of the electron's spin and the magnetic field inside the ferromagnet. The more statuses that are accessible, the more complex the probability of scattering, and the greater the electrical resistance. If the spin and magnetic field are anti-parallel, more circumstances are accessible for electron scattering, therefore the electrical resistance is larger than if the spin and the magnetic field are parallel. This is the basic idea of spin-dependent scattering [57]. In other words, the spin is a function of the inelastic scattering cross section (inverse mean free path) and the elastic scattering cross section are calculated for polarized electrons, which are scattered from oriented atoms in the Born-Ockhur approximation with a view of understanding spin-dependent scattering in ferromagnets. According to Matthew et al. [57], in the medium-to-high-energy range ($\gtrsim$ 100 eV) the elastic scattering for parallel spins is higher than for antiparallel spins, while the inelastic cross section for parallel spins is less than for antiparallel spins. Elastic spin dependence seems to be larger than inelastic, and the exchange effects fall off rapidly with increasing energy.

The discrepancy and the entire scattering cross-section could be considered as the experimental measures of the scattering challenge. The purpose of the scattering theory is then, to presume information on the force, or the interaction, responsible for the scattering. It is a crucial issue as radar object imagine, which is entirely based on the scattering theory. In this regard, the scattering theory achieves this challenge by establishing a relationship between the cross-section and the wavefunction of the radar system for instance.

4.4 Correlation between the Scattering Cross-section to the Wave Function

The expression "elastic scattering" infers that the internal circumstances of the scattering particles do not swap, and hereafter they develop unmovable from the scattering progression. In inelastic scattering, however, the particles' internal stage is revolutionized, which possibly will quantity to stimulating approximately of the electrons of a scattering atom, or the widespread annihilation of a scattering particle and the formation of exclusively different particles. In this view, the scattering problem is considered as time-independent and mainly is amplitude—dependent. Conversely, the time-independent Schrödinger equation of the scattering is expressed as [54–59]:

$$\left[U(r) - \hbar^2 (2\mu)^{-1} \Delta \right] \Psi(\vec{r}) = E \Psi(\vec{r}), \tag{4.3}$$

where

$$E \equiv E_k = \hbar^2 k^2 (2\mu)^{-1}, \tag{4.4}$$

Here $\Delta \equiv \nabla^2$ is the Laplacian operator, $\hbar \vec{k}$ is the momentum of incidents particles or photons. Equation 4.3 is assumed that the potential $U(r)$ is of finite range and it tends to zero faster than r^{-1} as $r \rightarrow \infty$. Equation 4.3 demonstrates that a homogenous beam of photons with momentum $\hbar \vec{k}$, which is directed in the positive z-direction and is given by:

$$\Psi_{incident} \equiv \Psi_k^i \underset{r \rightarrow \infty}{\sim} e^{\left(i \vec{k}.\vec{r} \right)} = e^{(ikz)} \tag{4.5}$$

Therefore, long after scattering, which signifies photons distributing in all directions from the scattering center. In this view, the flux of photons is denoted by a spherical wave. On the contrary, Equation 4.5 denotes the flux of photons which corresponds to plane waves (Fig. 4.3). Conversely, a plane wave is characterized by a definite linear momentum $\hbar \vec{k}$, but no definite angular momentum. In fact, a plane wave, being in principle, of infinite extension, corresponds to impact parameters varying from zero to infinity. Correspondingly, the angular momenta contained in a plane wave (x,y) also vary from zero to infinity. In this regard, it is possible, therefore, to analyze a plane wave into an infinite number of components each of which corresponds to a definite angular momentum. Each of such components is well known as a partial wave, and the process of decomposing a plane wave into the partial waves is called a partial wave analysis [52–58]. In this view, $e^{\left(i \vec{k}.\vec{r} \right)}$ is considered as a solution of the free-particle Schrödinger equation which defines the linear momentum representing the system as:

$$\hat{H} \Psi(\vec{r}) = E \Psi(\vec{r}) \tag{4.6}$$

with

$$\hat{H} = -\hbar^2 \Delta (2\mu)^{-1} \tag{4.6.1}$$

$$\Delta = \left[r^{-2} \frac{\partial}{\partial r}(r^2 \frac{\partial}{\partial r}) - \vec{L}^2 \left(\hbar^2 r^2 \right) \right] \tag{4.6.2}$$

where

$$\left[H, \vec{L}^2 \right] = 0, \tag{4.7}$$

$$\left[H, \hat{L}_z \right] = 0, \tag{4.8}$$

Here $\vec{L}^2$ and $\hat{L}_z$ are angular momentum operators and are given by:

$$\vec{L}^2 = -\hbar^2 \left[\sin^{-1}\theta \frac{\partial}{\partial \theta} \left(\sin\theta \frac{\partial}{\partial \theta} \right) + \sin^{-2}\theta \frac{\partial^2}{\partial \phi^2} \right] \tag{4.9}$$

$$\hat{L}_z = -i\hbar \frac{\partial}{\partial \phi} \tag{4.10}$$

Both Equations 4.7 and 4.8 imply the eigenvectors of $\hat{H}$, which are also the eigenvectors of $\vec{L}^2$ and $\hat{L}_z$. However, the eigenvectors of $\vec{L}^2$ and $\hat{L}_z$ are spherical harmonics $\gamma_{lm}(\theta, \phi)$ is formulated as:

$$\gamma_{lm}(\theta, \phi) = \left(-1^{l+m} \right) 2^{-l} l! \left[\frac{(2l+1)!}{4\pi} \frac{(l-m)!}{(l-m)!} \right] \sin^{|m|}\theta \times$$

$$\left(\frac{d}{d(\cos\theta)} \right)^{l+m} \sin^{2l}\theta . e^{im\phi} \tag{4.10}$$

$\gamma_{lm}(\theta, \phi)$ is called the spherical harmonic of order l. Further, $\gamma_{lm}(\theta, \phi)$ is the solution of the Laplace equation on the unit sphere ($r = 1$) hence the name spherical harmonics. m is $-l, -l + 1, \ldots, +l$. In this circumstance, l must be larger than m and must be a positive interger or zero, where $l = 0, 1, 2, 3, \ldots, +\infty$. In this understanding, $e^{\left(i\vec{k}.\vec{r} \right)}$ can be formulated in the form of outgoing spherical wave and scattering spherical waves, respectively as:

$$e^{\left(i\vec{k}.\vec{r} \right)} = 4\pi \sum_{l=0}^{\infty} \sum_{m=-l}^{+l} i^{-l} j_l(kr) \gamma^*_{lm}(\theta_k, \phi_k) \gamma_{lm}(\theta_r, \phi_r) \tag{4.11}$$

where (θ_k, ϕ_k) define the direction of $\vec{k}$ and (θ_r, ϕ_r) that of $\vec{r}$. $\dfrac{e^{-\left(i\vec{k}.\vec{r} \right)}}{r}$ represents the scattering spherical wave as $\vec{k}$ and $\vec{r}$ are antiparallel so that $\vec{k}.\vec{r} = kr\cos\pi = -kr$. However, $\dfrac{e^{\left(i\vec{k}.\vec{r} \right)}}{r}$ represents an incident spherical wave through radar antenna $\dfrac{e^{-\left(i\vec{k}.\vec{r} \right)}}{r} \vec{k}$ and $\vec{r}$ are parallel. Conversely, the wave function of the incident beams and scattering one can be formulated as:

$$\Psi_k(\vec{r}) \underset{r \to \infty}{\sim} 0.5 (ikr)^{-1} \sum_{l=0}^{\infty} (2l+1)[(-1)^{l+1} e^{(-ikr)} +$$

$$e^{(2i\hat{c}_l)} e^{(ikr)}] P_l(\cos\theta). \tag{4.12}$$

where P_l is the Legendre polynomial of order l. The first term of Equation 4.12 represents a scattering spherical wave while the second term represents the incident beam as a spherical wave, where both have the same intensity. Therefore, the phase of the incident wave is shifted relative to the phase of the corresponding wave in Equation 4.11. As a result, the spherical wave phase is shifted by the amount ∂_l, which is called the phase-shift [60–64].

Conversely, the mathematical description of spherical wave scattering is formulated as:

$$\Psi_{scattering} \equiv \Psi_k^S(r) \underset{r\to\infty}{\sim} f_k(\Omega)e^{(ikr)}r^{-1}. \tag{4.13}$$

where $f_k(\Omega)$ is known as the scattering amplitude. In this understanding, the scattering particles must be in the form of the spherical wave with the same energy as the incident one. In addition, Equation 4.6 demonstrates that for the great distances, it is the similar quantity of particles spreading through any cross-section of the given, solid angle element, as supposed (Fig. 4.4).

Equation 4.13 can be used to obtain the scattering cross-section based on Equations 4.1 and 4.2 as:

$$\frac{d\sigma}{d\Omega} = \left| f_k(\Omega) \right|^2 \tag{4.14}$$

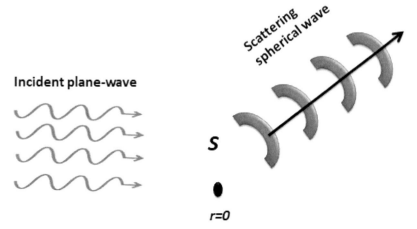

Figure 4.3. Mechanism of spherical wave scattering.

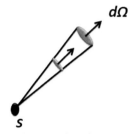

Figure 4.4. Cross-section of a given solid angle.

The scattering amplitude $\left| f_k\left(\Omega\right)\right|^2$ is independent of the azimuthal angle. In this regard, Equation 4.14 can be written as a function of Equation 4.12:

$$\frac{d\sigma}{d\Omega} = \left(k^{-2}\right)\sum_{l=0}^{\infty}\sum_{l^{'}=0}^{\infty}\left(2l+1\right)\left(2l^{'}+1\right)\times e^{\left[i\left(\partial_l-\partial_r\right)\right]}\sin\partial_l\sin\partial_r P_l\left(\cos\theta\right)P_{l^{'}} \tag{4.15}$$

Then the total cross-section is obtained from Equations 4.2 and 4.15 as:

$$\sigma = \int\frac{d\sigma}{d\Omega} = \frac{4\pi}{k^2}\sum_{l=0}^{\infty}\left(2l+1\right)\sin^2\partial_l = \sum_{l=0}^{\infty}\sigma^l \tag{4.16}$$

In Equation 4.15 $\displaystyle\sum_{l=0}^{\infty}\sigma^l$ presents the contribution to the total scattering cross-section from the partial wave which is constrained by

$$\sigma^l \le \left(\frac{4\pi}{k^2}\right)\left(2l+1\right) \tag{4.17}$$

From Equations 4.15 and 4.17, we have

$$\sigma^l = \left(\frac{4\pi}{k}\right)\text{Im}\left\{f_k\left(\theta=0\right)\right\} \tag{4.18}$$

where, $\text{Im}\left\{f_k\left(\theta=0\right)\right\}$ is the imaginary part of scattering amplitude. Equation 4.18 indicates that the total cross-section represents the loss of intensity suffered by the incident beam. In other words, some particles have been deflected away from the incident angle. Therefore, this loss of intensity is represented by the imaginary part of the scattering amplitude in the forward direction. Consistent with the above perspective, the thickness of the object has to be large enough to instigate adequate scattering energy, but small enough to retain multiple scattering at a minimum [54–60].

4.5 Scattering from Roughness Surface

Surface roughness scattering or interface roughness scattering is the elastic scattering of a charged particle by a defective interface between two different particles. It is a significant consequence in electronic devices which encompass narrow layers, for instance, field effect transistors and quantum cascade lasers.

Interface roughness scattering is most conspicuous in restricted systems, in which the energies for charge carriers are regulated by the settings of interfaces. An example of such a system is a quantum well, which may be created from a sandwich of different layers of the semiconductor. Differences in the thickness of these layers, therefore, triggers the energy of particles to be dependent on their in-plane location on the surface [60]. Though the roughness $\Delta_z(\vec{r})$ differs in an intricate approach on a microscopic scale, it can be reflected to exhibit a Gaussian Fourier transform $\Delta_z(\vec{r})$, which is considered by a height of Δ and correlation length ℓ. In this regard, the scattering from the roughness surface can be formulated as:

$$\left\langle \Delta_z(\vec{r})\Delta_z(\vec{r}\,')\right\rangle = \Delta^2 e^{\left(-\left|\vec{r}-\vec{r}\,'\right|^2\ell^{-2}\right)} \tag{4.19}$$

The frequently depleted expression for the ensuing scattering rate accepts a sudden interface geometry. This has been precisely fitted to investigation data for structures with the almost abrupt crossing-section, but Equation 4.19 is incompatible with smooth envelope potentials. In this regard, The perturbation $\Delta_V(\vec{R})$ due to a position shift $\partial_z(\vec{r})$ in an arbitrary confining potential U (z) is assumed to be correlated over the length of a single interface. At the point r = x + y assuming isotropy across the x and y-plane,

$$\Delta_V(\vec{R}) \sim -\partial_z(\vec{r})\frac{dU(z)}{dz} \tag{4.20}$$

Under time-independent perturbation theory, the perturbation must be small, i.e., $\partial_z(\vec{r}) \rightarrow 0$ and the perturbing potential simplifies to

$$\Delta_U(\vec{R}) = U_0(z_I)\partial_z(\vec{r})\partial(z-z_I) \tag{4.21}$$

where the *I*-th interface in a multilayer structure is centered on the plane $z = z_I$ and extends over the range $(z_{L,I}, z_{U,I})$. In this regard, the scattering matrix *S* can be formulated as:

$$S = \left\langle f \left| U_{0,I}\partial(z-z_I) \right| i \right\rangle \tag{4.22}$$

where $|f\rangle$ and $|i\rangle$ are the final and initial wave functions respectively.

4.6 Scattering of Identical Particles

Let us assume that the two particle states satisfy $\psi(1,2) = \psi(2,1)$. In this circumstance, a two-particle state ψ that is symmetric (antisymmetric) in r_1 and r_2 must be—an even (odd) function of the relative—coordinates $r \equiv r_1 - r_2$. However, in spherical coordinate $r \rightarrow -r$, which indicates that $(r,\theta,\phi) \rightarrow (r,\pi-\theta,\phi+\pi)$. In this regard, the wave function is neither symmetric nor antisymmetric. Consequently, for identical particles, it must be swapped with

$$\psi(1,2) = e^{i\vec{k}.\vec{r}} \pm e^{-i\vec{k}.\vec{r}} + \left[f(\theta) \pm f(\pi-\theta)\right]\frac{e^{i\vec{k}.\vec{r}}}{r} \tag{4.23}$$

Equation 4.23 demonstrates that the upper sign is expended for asymmetric wavefunction, and the lower for an antisymmetric wavefunction. In this regard, the spherical wave $\dfrac{e^{i\vec{k}.\vec{r}}}{r}$ is considered for the scattering of particles in diametrically opposite directions. From the point of view of physics, the scattering of both particles $\psi(1,2) = \psi(2,1)$ must be assumed when $\psi(1,2) = \psi(2,1)$ are matching. In this view, abrupt change of the scattering cross-section $d\sigma$ is identified by the ratio of the particle flux into $d\Omega$ (Equation 4.2) and the incident particle stream for one of two plane waves (x,y). This can be formulated mathematically as:

$$\frac{d\sigma}{d\Omega} = \left| f(\theta) \pm f(\pi - \theta) \right|^2 \tag{4.24}$$

Equation 4.24 is consistent with quantum mechanical treatment. In fact, this can be done by adding a wavefunction and then computing the probabilities by taking the absolute value squared. This leads to two possibilities: (i) scattering of particles with spin, and (ii) scattering of spin-0 particles $d_\Omega\sigma = \left| f(\theta) \right|^2 + \left| f(\pi - \theta) \right|^2$, which has no interference term between $f(\theta)$ and $f(\pi - \theta)$ [53–63].

4.7 Scattering of Particles with Spin

Let us take into account the fact that the corrected $d_\Omega\sigma$ is obtained by the interaction between dual spinful particles. This can be proved by assuming that $e - e$ scattering (spin 1/2). In this case, dual spin 1/2 states may be combined into one singlet $(\uparrow\downarrow - \downarrow\uparrow)$ or three triplet $(\uparrow\uparrow, \downarrow\downarrow, \uparrow\downarrow + \downarrow\uparrow)$ states. The probability of 0.25 for singlet and probability ¾ for triplet are considered when the particles are randomly polarized. In this circumstance, the corrected $d_\Omega\sigma$ is given by:

$$d_\Omega\sigma = \frac{3}{4} \left| f(\theta) \right|^2 + \left| f(\pi - \theta) \right|^2 + \frac{1}{4} \left| f(\theta) + f(\pi - \theta) \right|^2 \tag{4.25}$$

Equation 4.26 infers that particles scattered into $\theta = 0.5\pi$, which must be a singlet. However, the triplet influence is zero for this θ. Furthermore, if the spins are not primarily arbitrary, however, fully polarized in the similar θ, the scattering into $\theta = 0.5\pi$ is zero [58–62].

4.8 Scattering of Zero Spin Particles

In the case of the spin-0 particle, the bosons are considered as a symmetry wavefunction. Let assume for concreteness that the bosons intermingle through the Coulomb-potential, which can be formulated as:

$$f_c(\theta) = \frac{Z^2 m}{k a_0 m_e} \left(2k \sin^2(0.5\theta) \right)^{-1} e^{-2i\ln\sin(0.5\theta) + i\partial} \tag{4.26}$$

Use Equation 4.24 with Equation 4.26, abrupt change of scattering cross-section $d\sigma$ is given as:

$$\frac{d\sigma}{d\Omega} = \left(\frac{Z^2 e^2}{4\pi\varepsilon_0 E_T} \right) \left[\sin^{-4}(0.5\theta) + \cos^{-4}(0.5\theta) \right] +$$

$$\frac{2\cos\left[\frac{Z^2 m}{k a_0 m_e} \ln\tan^2(0.5\theta) \right]}{\sin^2(0.5\theta)\cos^2(0.5\theta)} \tag{4.27}$$

where $E_T = \dfrac{\hbar^2 k^2}{2m}$ as m is the electron's mass. The last term in Equation 4.27 is a purely quantum mechanisms influence slowing from the interference between $f(\theta)$ and $f(\pi - \theta)$ [58–63].

4.9 Resonant Scattering

From the point of view of physics, resonant scattering is defined as phenomena of scattering which is occurring when an electromagnetic wave hits an atom and then is scattered in another direction (Fig. 4.5). In other words, if a photon's power suits (almost) precisely the intensity distinction between the ground phase and the lowest ("first") excited phase of an atom, then when the atom de-excites, every other photon with (almost) precisely the identical power is emitted, however in every other direction. This is known as resonant scattering, due to the fact the photons are in resonance with the atomic transition or the line. If the photon has a slightly too short or long wavelength, it will genuinely pass right through the atom [52–55].

Conversely, the photon's wavelength can gradually exchange. In this understanding, a photon with a shorter wavelength is scattered, when the atom is transferred away from the incident photon with a velocity U. In this sense, the Doppler shift has swung the wavelength of the photon close to resonance, for instance, in the reference frame of the atom. In this view, when the direction is opposite U, the photon will appear to be radiated with a longer wavelength [57].

To understand the resonant scattering, let us assume low-energy scattering on a well-potential:

$$P_0(r) = \begin{cases} -P_0 \text{ for } r \leq R \\ 0 \text{ for } r > R \end{cases} \tag{4.28}$$

In this regard, the low energy $kR \ll 1$ and only the partial wave l contributes to the cross-section according to:

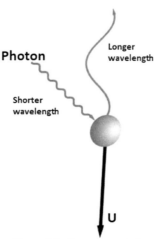

Figure 4.5. Resonant scattering.

$$\sigma = 4\pi k^{-2} \sin^2 \hat{\partial}_0 \tag{4.29}$$

The total cross-section is expressed as:

$$\sigma = 4\pi \left(k^2 + \varsigma^2 \right)^{-1} \tag{4.30}$$

where

$$\varsigma = \kappa (\tan^{-1}(\kappa R))^{-1} \tag{4.30.1}$$

Equation 4.30, indicates that for a given k, the cross-section of resonant scattering is maximal when $\varsigma = 0$. The mathematical conversation expression of the resonant energy is given as:

$$E = -P_0 + \hbar^2 \pi^2 (2mR^2)^{-1} [n + 0.5]^2 \tag{4.31}$$

where n is an integer, whereas Equation 4.31 indicates that when the incident particle has just the accurate resonant energy, Equation 4.31 will have a propensity to be assured by the potential and remain at $r \leq R$. In this circumstance, a major disturbance of the wavefunction is produced and causes a large scattering cross-section σ. This resonance scattering is the keystone to understand the mechanism of synthetic aperture radar for imagining sea surface [54–60].

The significant question that can be raised up is "what is resonance Rayleigh scattering and how is different from dynamic photon scattering?

Resonance Rayleigh Scattering is a phenomenon rather than a technique. When a tiny particle/molecule has an absorption line, the quantity of scattered photon significantly growths at wavelengths near the absorption line owing to a modification in the complex refractive index of the particle/molecule. Resonance light scattering techniques frequently exploit fluorometers scanning with the same excitation and emission wavelengths to recognize these scattering peaks. If performed properly then quantitative consequences can be acquired. Therefore, dynamic photons scattering determines the hydrodynamic size of particles, which it achieves by quantifying the variability in photon scattered by the particles. The faster the fluctuations, the smaller the particles [65].

4.10 Dielectric Materials and Electric Polarizability

In electronic transmission, either the free motion of the electrons or combined diffusion of the ions is expected. For instance, when the charged particles travel, a current is generated. However, in dielectric materials, electric dipoles play a dominant role in the properties of the material instead of the motion of the electrons or ions.

The mechanism of polarizability that instigates the microwave absorption contains rotation and orientation of the dipoles [66]. Conversely, there are three approaches this can solid materials:

i) The single atom can have the form of the "electron cloud" in closing the nucleus twisted by the electric field. Generally, atoms with numerous electrons, i.e., a high atomic number are very simply deformed and are deliberated exceptionally "polarizable."

ii) Molecules with a stable electrical dipole have the dipole affiliated in reaction to the electric field.

iii) Molecules, with or without a permanent dipole, can have bonds deformed, i.e., direction and length in effect of an electric field.

Deprived of an electric field, the dipoles are concerned with moment and the net polarization is zero. However, in a static field, the dipoles associated with the field and the polarization is maximized. These dipoles can revolve, and then they swap against a friction force. When the frequency of the electric field grows, the revolution of the dipoles cannot be monitored, and the net polarization in the material is no longer in phase with the electric field.

Under an electric field, E, the electron cloud in an atom may be expatriated compared to the nucleus, separation an uncompensated charge-q at one side of the atom and +q at the other side. The uncompensated charges, therefore, create an electric dipole moment and the sum of those dipoles over a unit volume is the polarization. Similarly, deformation of a charged ion relative to other ions produces dipole moments in the molecule. Those are the atomic and ionic polarizations produced by the electric field. Though atomic and ionic polarizations ensue at microwave frequencies, they do not cause the heating of the microwave.

Pure water is a nonpolar dielectric. Water molecules are moving randomly. They are not tightly bound to each other. In an electric field, the water molecules are polarized. But they are not at rest and cannot produce charges to create an electric field like a solid dielectric. The motion of water molecules differs from the function of a capacitance constantly. Therefore, water cannot be used as a dielectric in a capacitor. In this regard, when the seawaters are a perfect conductor, the incident field would create surface charges and currents that would precisely abandon the fields in the inner of the surface. This is not the case, but it is a pretty close—reflectance, that is more than 60%, for instance. Thus, the imaginary part of the dielectric constant, closely related to conductivity, makes the ocean behave like an attractive excellent mirror at microwave frequencies, which does not occur at optical frequencies. In other words, the microwave frequencies cannot penetrate the sea surface due to its extremely s high dielectric [66,67].

4.11 Atom-Photon Scattering

Let us assume that an atom O_i interacts with a photon γ_i, which leads to the atom O_j and a photon γ_f in probably diverse quantum states. In this regard, the atomic scattering of photon radiation can be formulated as:

$$O_i + \gamma_i \rightarrow O_f + \gamma_f \tag{4.32}$$

Equation 4.32, indicates that the scattering occurs in different quantum energies, which can mathematically be written as:

$$E_i + \hbar\omega_i = E_f + \hbar\omega_f \tag{4.33}$$

Equation 4.33 is conversion quantum energy states, where E indicates the atomic energies and $\hbar\omega$ refers to the photon energy. Indeed, the incident photon energy interaction with atomic energy is equivalent to the scattering energy of photon due to different atomic energies. Conversely, there are four sorts of atom-photon scattering procedures: (i) Rayleigh Scattering; (ii) Raman Scattering; (iii) Thomson Scattering; and (iv) Compton Scattering [68]. Both Rayleigh Scattering and Raman Scattering present low-energy elastic scattering and inelastic scattering, where $\omega_i = \omega_f$ and $\omega_i \neq \omega_f$, respectively. However, both Thomson Scattering; and Compton Scattering present high-energy elastic scattering and inelastic scattering, where $\omega_i = \omega_f$ and $\omega_i \neq \omega_f$, respectively. In this view, microwave remote sensing is concerned with the low-energy photon scattering.

At the glummest imperative in non-relativistic perturbation speculation, two possible Feynman diagrams can explain the process of photon scattering by an atom (Fig. 4.6). Nevertheless, in relativistic quantum field theory, both time-ordered diagrams would be signified by a single Feynman diagram. In this understanding, since the photon is always a relativistic particle, the atom must be chosen as in nonrelativistic kinematic phases. As all the particle travels in the relevant diagrams denote to atoms, non-relativistic perturbation theory can be used to designate the dynamics of the atom.

Phase I demonstrates that the atom O_f absorbs the photon γ_i transforming into the intermediate O_n formal. Subsequently, the atom emanates a photon γ_f and transfers into the O_f status. The mathematical expression of these circumstances is given by:

$$O_i + \gamma_i \rightarrow O_n \rightarrow O_f + \gamma_f \tag{4.34}$$

Phase II presents that the atom O_i emanates the photon γ_f and changes into the intermediate state of O_n. Successively, the atom absorbs the photon γ_i and enters into the O_f status. This circumstance can be expressed mathematically as:

$$O_i + \gamma_i \rightarrow O_n + \gamma_f \rightarrow O_f + \gamma_f \tag{4.35}$$

Let us assume that absorption or emission of photons can be symbolized through four states:

i) $V_{ni}^a : O_i$ absorbs a photon ω_i and transforms into O_n.
ii) $V_{fn}^e : O_n$ emanates a photon ω_f and transforms into O_f.
iii) $V_{ni}^e : O_i$ radiates a photon ω_f and changes into O_n.
   i) $V_{fn}^a : O_n$ absorbs a photon ω_i and converts into O_f.

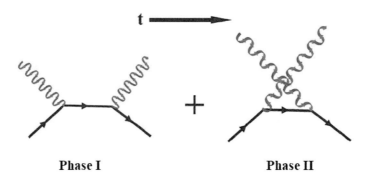

Figure 4.6. Feynman diagrams of photon scattering.

In terms of atomic and photonic quantum states, the mathematical descriptions of the above four states can be given by:

$$V_{ni}^a = \left\langle O_n \middle| \otimes \left\langle 0 \middle| \hat{V}^a \middle| O_i \right\rangle \otimes \middle| \omega_i \right\rangle$$

$$V_{fn}^e = \left\langle O_f \middle| \otimes \left\langle \omega_f \middle| \hat{V}^e \middle| O_f \right\rangle \otimes \middle| 0 \right\rangle$$

$$V_{ni}^e = \left\langle O_n \middle| \otimes \left\langle \omega_f \middle| \hat{V}^e \middle| O_i \right\rangle \otimes \middle| 0 \right\rangle \qquad (4.36)$$

$$V_{fn}^a = \left\langle O_f \middle| \otimes \left\langle 0 \middle| \hat{V}^a \middle| O_n \right\rangle \otimes \middle| \omega_i \right\rangle$$

here $\hat{V}^a$ and $\hat{V}^e$ are quantum operators, which signify the atomic absorption and emission of a photon, respectively. Consequently, non-relativistic perturbation theory stipulates the transition amplitude for the atom-photon scattering process is given by:

$$\tilde{V}_{fi} = \sum_n \left(\frac{V_{fn}^e V_{ni}^a}{\left(E_i + \hbar\omega_i \right) - E_n} + \frac{V_{fn}^a V_{ni}^e}{E_i - \left(E_n + \hbar\omega_f \right)} \right) \qquad (4.37)$$

The energy states of E in Equation 4.37 explained that in minimal coupling quantum electrodynamic (QED), the interaction between a photon of low energies and matter is specified by a term in the energy that couples a charged particle field with the quantum electromagnetic field [69,70].

4.12 Quantum Young Scattering

Young's experiment is a demonstration that electromagnetic wave and matter can reveal physical characteristics of both typically definite waves and particles; likewise, it shows the primarily probabilistic nature of quantum mechanical phenomena. In Young's experiment, a coherent electromagnetic source, for instance, laser beam, illuminates an electroplate stabbed by dual-parallel slits. Consequently, the coherent light spreading through the slits, which is detected on a screen behind the electroplate. Conversely, the coherent light interferes and generates, producing bright and dark bands on the screen—a concern that would not be presumed if light involved particles. From the point view of quantum mechanics, let us assume that the scattering occurs due to the photon interacting with ground b_i and energized E states of atom i atom in time t. In fact, photons have a certain probability to interact with an atom, which depends on the energy at a given physical location (which depends on waves superposition) and on the atoms used to construct the sensor (should not be transparent at the wavelength used by the coherent source).

The photon is captivated by one of the atoms, which is directed from the ground to the energized state. Successively, the atom radiates the photon and turn back to its ground state. The quantum circumstances of the array before, after and during the photon have been captivated are formulated as:

$$\left| \Psi \right\rangle = \frac{1}{\sqrt{2}} \left| 0 \right\rangle \left(\left| e_1 b_2 \right\rangle + \left| b_1 e_2 \right\rangle \right) \qquad (4.38)$$

Equation 4.38 reveals that the photon interacts with both scattering atoms in a quantum superposition. However, the Young's experiment cannot determine the atom that scattered the photon, and such ambiguity creates the interference pattern appraised in the screen. In this circumstance, let us assume that Δr_{id} is the distance from the ith atom to the point r in the detecting screen and ΔR_i is the total interferometric distance of the experiment which is given by:

$$\Delta R_i = |r_s - r_i| + |r_i - r_d| \tag{4.39}$$

where r_s, r_d, and r_i are the locations of the coherent light source, the detecting point and the ith atom, respectively. The quantum explanation of the photon interferometric scattering can mathematically be written as:

$$\psi_\gamma^i(\Delta R_i, t) = \frac{\varepsilon_0}{\Delta r_{id}} \Theta\left(\frac{t - \Delta R_i}{c}\right) e^{-(i\omega + 0.5\Gamma)\left(\frac{t - \Delta R_i}{c}\right)} \tag{4.40}$$

where Γ is an attenuation rate. The alteration of the space of the coherent light source for every single of the scatterers influences the interference pattern. Consequently, the exponential and the step function necessitate the additional phase factor to offset for the accurate geometry of the scatterers. Alternatively, the amplitude term specifies the isotropic decay of the amplitude from the scatterer to the screen, which does not prerequisite to be revised by the geometry of the scatterers.

4.13 Can Describe Specular Reflections as a Scattering Process?

It is a significant question to understand, how the oil spill later can cause a specular reflection in the radar image. Let us assume that there are a number of atoms, covering the surface. These atom particles produce a mirror surface. Conversely, the photons interact with the atoms as explained previously, which causes the photon scattering. The photon scattering is due to an interference pattern which is analogous to the one acquired by Young's experiment with numerous scattering atoms. In other words, the incident photon concurrently intermingles entirely with the atoms in the surface, which are acting as the mirror. This influence is termed by a quantum superposition analogous to the one depleted in the investigation of Young's interference experiment with numerous scattering atoms. Certainly, the photon is scattered by the atoms and the power of the photonic electromagnetic quantum field is quantified by the sensor. Nevertheless, there is ambiguity in the definite atom that scattered the photon. Such a vagueness eventually precedes to the interference pattern determined by the sensor.

In general, photons obeying a wave-like equation grow and interfere in this circumstance identically corresponding to conventional waves, and that photon squared amplitudes provide a probability of identifying them—is certainly reliable with quantum field theory. In this context, the quantum field hypothetical, quantum mechanical and classical predictions for what the states of the quantum electromagnetic field, the photons, and the classical electromagnetic field do at the double slit are reliable.

Quantization of Radar Theory

Radar is an acronym for Radio Detection And Ranging. Radar was initially elaborated in place of technology for sensing targets and revealing their locations by means of echolocation, and thus persevered the primary function of modern radar systems. Nevertheless, radar systems have developed over more than seven decades to achieve a diversity of the precise complex functions; for instance, imaging is one such function. In this regard, radar imaging is a technology that has been generally established within the engineering community. There are noble causes for this: some of the precarious constituents are: (i) transmitting microwave energy at high power, (ii) detecting microwave energy, and (iii) interpreting and acquiring knowledge from the received signals. Radar imaging is a function of marvelous mathematical abundance. Particularly, it contains partial differential equations, scattering theory, microlocal analysis, integral geometry, linear algebra, electromagnetic theory, harmonic analysis, approximation theory, group theory and statistics. Conversely, radar can deliver precise distance (range) quantities and can also determine the rate at which this range is fluctuating. In this context, radar waves scatter principally from targets and structures whose sizes are in a similar order as the wavelength. Consequently, radar is extremely sensitive to targets whose length scales range from centimeters to meters, and perhaps objects of interest are in this range.

5.1 RAdio Detecting and Ranging

With intending to comprehend the mechanism of Synthetic Aperture Radar (SAR) data image of the oil slicks, the main characteristics of SAR must be fully understood. The active microwave Earth observation is also known as RADAR (RAdio Detection And Ranging). The term radio used because the first radar used long wavelengths of radiation (1 to 10 m) that fell on the radio band of the electromagnetic spectrum. The term radar is commonly used for all active microwave systems [21]. In this view, radar is defined as a system for detecting the presence, direction, distance and speed of aircraft, ships and other objects, by sending out pulses of radio waves which reflect off the object back to the source. In other words, radar is an acronym for "radio detection and ranging." A radar device commonly operates in the Ultra-High-Frequency (UHF) or microwave phase of the Radio-Frequency (RF) spectrum and is depleted to detect the locus and/or movement of targets [21,71].

The radar configuration consists of an antenna, transmitter and receiver. The transmitter generates the energy to provide a radar beam and transmits it to the antenna and therefore to the target. The transmitted signal is a short burst rapidly repeated to give the pulsed signal. These pulsed signals traveling at 300 000 km/s strike the targets in view where some of the energy is absorbed, some is reflected away and some refracted or backscattered to the receiver [72]. The receiver is usually integrated with the transmitter. The receiver gets the pulsed returned signal. It determines the signal strength and relates the signal to the transmission. To calculate target distance, these data are processed into a form suitable for recording. The return signal, consequently, is much lower than the transmitted signal since not all the signal is returned. This signal will provide the gray scale tone of the image [73].

This sort of microwave imaging is termed active because they transmit their own microwave energy (pulses) at a particular wavelength (single frequency) for a particular duration of time, known as pulsed coherent radar. Figure 5.1 lists the common concept of radar system. In this context, the following sections are concerned with the fundamental concepts of SAR image data [74].

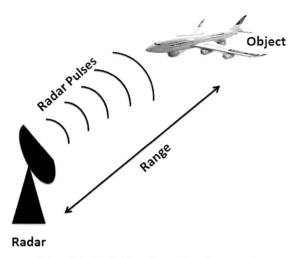

Figure 5.1. RAdio Detection and Ranging concept.

5.2 Echo-location Detecting and Ranging

Active radar systems succeed echolocation in conveying an electromagnetic wave and quantifying the reflected field as a time-varying voltage in the radar receiver. To comprehend the elementary thought of echolocation (Fig. 5.2), let us consider an ultimate radar array that launches, at period $t = 0$, a short pulse that propagates at the speed of light c, imitates from a target at a range R, and incomes to the radar. If the income pulse can be identified at a time τ (Fig. 5.3), then because $c\tau = 2R$ (speed $\times$

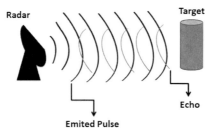

Figure 5.2. Echolocation concept.

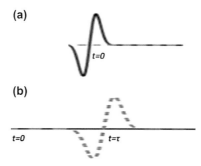

Figure 5.3. A pulse is transmitted at (a) time zero and (b) received on time τ.

time = round-trip distance), it can find that the range *R* must be equal to $R = c\tau/2$. In this view, it is considered as the basic theory of radar system.

5.3 High-Range-Resolution (HRR) Imaging

High-Range-Resolution (HRR) imaging is urgently required. In fact, most objects of interest are challenged to be individually detected. For instance, a superposition of reflections is acquired when the short pulse hits the airplane. Consequently, this response is well known as a high-range-resolution (HRR) profile, which can be deliberated as a one-dimensional "image" of the object (Fig. 5.4).

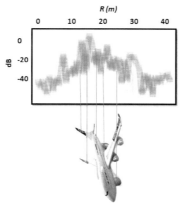

Figure 5.4. Radar reflections from a complex target.

5.4 Radar Microwave Characteristics

The microwave is part of the electromagnetic spectrum, which ranges from the wavelength of 1 mm to 1 m, corresponding to signal frequencies f equals 300 GHz and 300 MHz ($\lambda f = c$, with the speed of light c), respectively. Unlike the visible spectrum, the wavelength is extremely larger and intermingles with reasonably different targets contrasted to passive spectra. In this view, the microwave signal has lower energy than passive spectra to induce molecular resonance. In contrast, passive spectra of visible and near infrared, for instance, exposes their signal patterns into the atmosphere and soil chemical structures due to their high energy. The microwave domain, conversely, still have adequate high energy to model resonant spin of definite dipole molecules consistent with the frequency which is a function of the changing of a signal electric field [72].

In this sense, the object appears dark because it is smaller than the radar wavelength which does not imitate abundant energy. On the contrary, short-wavelength radar can discriminate minor variants of irregularity than long-wavelength radar. In other words, object roughness fluctuates with wavelength. Therefore, because of the diffuse reflection, the incidence angle does not have a significant role in rough surface [21].

SAR sensors are reasonably delicate to the object's physical properties, for instance, permittivity ε, surface roughness, morphology and geometry. Active microwave wavelengths tend to be divided into typical wavelength regions or bands (Table 5.1) (Fig. 5.5).

In microwave sensors, therefore, frequency (in Hertz) can also be used to describe a band range. It is worth mentioning that satellite active microwave sensors do not acquire multispectral microwave images. Conversely, they attain data for only a single band of wavelength/frequency. Nevertheless, recent airborne sensors can acquire multi-frequency bands, for instance, NASA's Jet Propulsion Laboratory AIRSAR system; C, L, and P-band.

Table 5.1. Microwave bands and their physical characteristics [21].

Bands	Wavelength (cm)	Frequency (GHz)
P-band	30–100	1.0–0.3
L-band	15–30	2.0–1.0
S-band	7.50–15	4.0–2.0
C-band	3.9–7.50	4.0–8.0
X-band	2.40–3.75	8.0–12.5
K-band	0.75–2.40	12.5–40 (rarely employed)

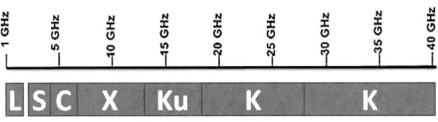

Figure 5.5. Microwave band spectra.

In addition, a theoretical quantum radar operating in the 9 GHz regime, which is able to improve object detection under the impact of atmospheric attenuation. In this regard, the photons in the 9 GHz frequency (Table 5.1) correspond to the radar-microwave spectrum. In this sense, the X band in radar corresponds to the 8–12 GHz region and is extensively used for missile guidance, marine radar, weather, ground surveillance and airport traffic control [76–78].

A quantum radar, consequently, functioning in X- the band would be tremendously valuable in a diversity of operational environments of national and military concern.

5.5 Why Quantum Radar Sensors are Required?

It is a important question to be answered. Indeed, quantum theories are required to increase the performance of a broad diversity of conventional information processing sensors. A talent quantum computing can be achieved by integration between quantum communication and advanced computer infrastructure systems. Further, the interface between quantum information science and quantum sensing is significant; for instance, quantum sensors can be designated as noisy quantum modes. Furthermore, quantum computation performances established in the perspective of quantum control, which is valuable to exploit quantum sensing hardware.

It is imperative to state that although quantum sensing is not established yet as quantum computation, it presents some manufacturing difficulties. As a quantum computation necessitates a huge quantity of qubits on uninformed superpositions with adequate coherence times to achieve convoluted computations exploiting a great diversity of accesses. In contrast, quantum sensors necessitate a minor integer of qubits on a precise entangled formal. As a result, it merely necessitates a minority of quantum processes. In consequence, it seems that the expansion of quantum sensors delivers a tantalizing near tenure possibility for the realistic use of the quantum information technologies demanded for the comprehension of a speculative quantum processor.

By developing entanglement, quantum radar provides the expectation of improved target recognition proficiencies. In this regard, the quantum radar relies on quantum disarrays of electromagnetic waves, i.e., microwave photons, which are maintained on a twisted superposition. In this understanding, it is directed towards the object and the backscattered signals in the receiver. In this view, signal detection is improved by utilizing the obtainable connection between the radiated microwave photons leaped back from the object and the ones coded into the radar. Impartially, quantum radar provides the option of distinguishing, recognizing and determining surreptitiousness objects. Consistent with Lloyd [75], over non-entangled photons, a quadratic resolution enhancement can improve using entangled photons through quantum radar. In this understanding, the visibility of target detection can be increased using a quantum radar. Indeed, a quantum sidelobe structure provides a new mode for the revealing of RF surreptitiousness objects.

5.6 What Does Mean by Quantum Radar?

Generally, a quantum radar can be defined as an impasse detection device that exploits microwave photons and uses some formula of quantum phenomena to improve its proficiencies to distinguish, recognize and determine an interesting object. In the case of this chapter, quantum radar is defined as the quantum phenomena which used to understand the functionality of the radar mechanism.

The objects are expected to retain a squat backscatter, which are far-off from the sensor. Moreover, the radar-target system is assumed to be engrossed in a noisy and lossy environment. Specifically, the radar signal may well be diminished by absorption or scattering developments, and the general implementation of the system is pretentious by the existence of noise.

5.7 What are the Classifications of Quantum Radar?

The main illustrations of quantum radar sensors are the single photon quantum radar and the entangled photon quantum radar. In other words, three sorts of quantum radar can be categorized as follows:

i) The quantum radar conveys un-entangled quantum circumstances of microwave photons.
ii) The quantum device radiates conventional statuses of microwave photons, however, exploits quantum photo-sensors to enhance its operation.
iii) The quantum sensor emits quantum signal states of microwave photons that are entwined with quantum ancilla statuses of a microwave photon generated at the transmitter.

The basic concept of a single photon quantum radar: a single photon is emanated towards a target, and consequently the photon is scattered back to the receiver (Fig. 5.6). In contrast, the mechanism of entangled-photon quantum radar, an entangled pair of microwave photons is produced. In this case, the photons are emitted towards the objects and the other stored in the radar sensor. The outward photon is, then reflected by the object and consequently received by the radar system. In this view, the connections entrenched in the entangled condition, which are oppressed to enhance detection operation (Fig. 5.7) [80].

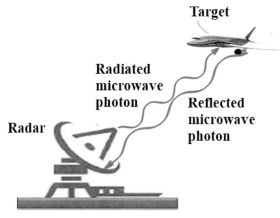

Figure 5.6. Quantum radar.

One of the quantum radar mode is LADAR, which stands for Laser Detection And Ranging, and it is has a similar technology to LIDAR (Light Detection And Ranging) (Fig. 5.8). Moreover, it operates in the visible and near-visible photon spectra [79]. However, it cannot be counted as a form of quantum radar technology. In fact, it cannot infiltrate fog or clouds as radars do, and as a consequence, the operational range of a LADAR is regularly constrained within 100 km [21].

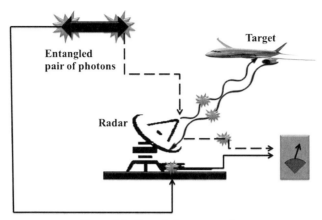

Figure 5.7. Entangled quantum radar.

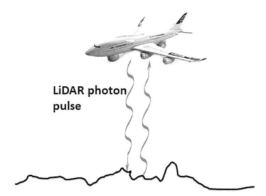

Figure 5.8. LiDAR photon pulse.

5.8 Classical Radar Equation

Radar range equation is the simplest mathematical description of a radar principle. Though it is one of the most effective equations, paradoxically, it is an equation that few radar analysts comprehend and plenty of radar analysts mishandle. The trouble, therefore, lies not with the equation itself, but with the numerous terms that form the equation. A deep understanding of the radar range equation delivers a completely solid basis of radar principle. The radar equation is the termination of several simpler formulas [21].

The transmitted energy density E_T can be explained as energy per unit area and occurs in a range R. The scientific explanation of the power density at a distance of the emitted signal can mathematically be written as:

$$E_T = \frac{P_T . \tau . G_T}{4\pi R^2} \tag{5.1}$$

In Equation 5.1, $4\pi R^2$ associates the power transmitted by the radar to an isotropic sphere. In this sense, the electromagnetic energy transmits in the same way in all directions (Fig. 5.9).

Further, G_T is the antenna and the focus of the antenna signal is a function of the ratio of $\frac{G_T}{4\pi R^2}$. Consequently, P_T is called the *peak transmit power* and is the average power when the radar is transmitting a signal. P_T can be specified on the output of the transmitter with a duration time τ of or at some other point like the output of the antenna feed. It has the unit of watts.

Equation 5.1 can be developed as a function of the Radar Cross Section (RCS) σ which is a result of the backscatter size of the target on which the radar signal is focused (Fig. 5.10). Therefore, the radiated energy of an object is estimated via:

$$E_\sigma = \frac{P_T . \tau . G_T . \sigma}{4\pi R^2} \tag{5.2}$$

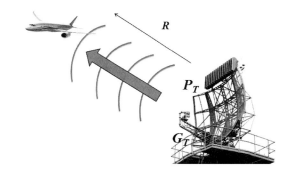

Figure 5.9. Transmitted pulse from radar antenna.

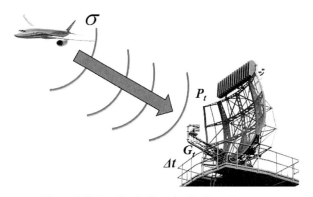

Figure 5.10. Received reflected pulse from the target.

Equation 5.2 expresses the backscatter power density as a function of Radar Cross Ssection (RCS). Therefore, RCS relies on the object's unique scattering characteristics. Further, the target *radar cross-section or RCS* has the units of square meters or m².

The energy of the receiving antenna signal S is a function of the backscatter and is located at the same transmit antenna. This is known as monostatic radar. In this view, the term of $4\pi R^2$ in the previous equation turns into $(4\pi)^2 R^4$ with the additional parameter of A_R in the numerator. The term A_R, conversely, is the operational area of the receiving antenna and is the numerator in a ratio concerning the second $4\pi R^2$. This represents the isotropic radiation owing to a target's radar cross-section [71–77]. Under this circumstance, Equation 5.2 can be modified as:

$$S = \frac{P_T . \tau . G_T . \sigma . A_r}{(4\pi)^2 R^4} \tag{5.3}$$

Equation 5.3 articulates the quantity of signal which reaches on the receiving antenna. Noise, therefore, can be generated in the receiver of the radar system. This noise is known as the Signal-to-Noise Ratio (*SNR*) and has the units of watts/watt, or w/w. Further, the radar system is also dominated by additional factors such as thermal noise temperature T_0, Boltzmann's constant K, and is equal to 1.38×10^{23} w/(Hz °K). The receiver bandwidth B, and L represent losses within the system itself. One last element which needs to be developed for Equation 5.3 is a relation between antenna gain, its effective area, and signal wavelength λ.

$$A_R = \frac{G_R \lambda^2}{4\pi} \tag{5.3.1}$$

Equation 5.3.1 expresses the correlation between antenna area and gain, i.e., gain and effective aperture relation. The radar final formula results from substituting A_R and the noise constants into the previously developed formulas. In this view, the scientific explanation of radar range can mathematically be written as:

$$SNR = \frac{P_S}{P_N} = \frac{P_T G_T G_R \lambda^2 \sigma}{(4\pi)^3 R^4 k T_0 B F_n L} \tag{5.4}$$

where, F_n is the radar *noise figure* and is dimensionless, or has the units of w/w, and L is a term involving all losses that must be considered when using the radar range equation. L has the units of w/w. Equation 5.4 is the last version of the radar equation. Additional terms can be added or substituted for diverse schemes. For instance, occasionally, the transmission interval denotes the combination of countless signals that occur over the total time t. The main concept to consider from the radar equation is the inter-need of various diverse aspects of radar [21,76].

5.9 Quantum Radar Equation

Quantum radar equation is a function of radar cross-section. In this sense, how do we define a quantum radar cross-section? Let us consider that σ_Q is quantum radar cross-section, which is mathematically defined as [80]:

$$\sigma_Q = \lim_{R \to \infty} 4\pi R^2 \frac{\langle P_S \rangle}{\langle P_N \rangle} \tag{5.5}$$

The transmitted power P_S is formulated as:

$$P_S \approx \frac{4\pi \varepsilon_0^2 \sigma_Q}{(4\pi)^2 R^4} \tag{5.6}$$

Equation 5.6 shows the classical radar equation. However, the quantum radar equation which involves the transmitted P_T^Q and reflected power P_r^Q, respectively, and can be formulated as:

$$P_T^Q = 4\pi \varepsilon_0^2 \tag{5.7}$$

$$P_r^Q = \langle P_S \rangle A_R \tag{5.8}$$

The quantum radar equation can be formulated using Equations 5.7 and 5.8 as:

$$P_r^Q = \frac{P_T^Q A_r \sigma_Q}{(4\pi)^2 R^4} \tag{5.9}$$

σ_Q has the following significant properties:

Strong Dependencies: It can be proved that σ_Q strongly depends on the properties of the target. That is, it depends on the target's geometry (absolute and relative size, shape, and orientation), as well as its composition (material properties).

Weak Dependencies: It can also be proved that σ_Q is approximately independent of the properties of the radar system. That is, it depends very weakly on the strength and the position of the radar system. In particular, σ_Q is independent of R, the range to the target. As a consequence, it is clear that σ_Q is a property that (approximately) characterizes a specific target, and not the radar system and/or its interaction with the target. In this case, the simulation of σ_Q for the proposed design of a vehicle will provide a good estimate of its "radar invisibility". In addition, σ_Q is also important to characterize the operational performance and capabilities of radar systems. That is, given a radar system, what is the minimum σ_Q of a target that it can detect [21,67].

5.10 Radar Scattering Regime

Depending on the value of the ratio of the characteristic size of the target O and the wavelength of the radar λ, three scattering regimes characterize the operation of radar [76–80].

Rayleigh Regime $O \ll \lambda$: This is the low-frequency case, and the incident electromagnetic wave shows little phase variation over the physical extent of the target. That is, at each instant of time, each portion of the target is affected by nearly the same electromagnetic field. In a sense, at each moment of time, the electromagnetic field is approximately constant over the entire physical extension of the target. Then, the incident electromagnetic wave induces dipole moments which only depend on the size and orientation of the target.

Resonant Regime ($O \approx \lambda$): In this case, the phase of the incident electromagnetic field changes along the length of the target. This means that some of the energy of the electromagnetic field is "attached" to the target's surface, creating surface waves that travel over the spatial extent of the target.

Optical Regime $O \gg \lambda$ In the high-frequency case, collective interactions are minimal and the body can be considered made of a collection of independent scattering centers. The total electromagnetic scattered field is the superposition of all the individual scattered fields. As a consequence, the geometry of the target plays a dominant role in the structure of the scattered fields [77–80].

5.11 Quantum Theory of Radar System

5.11.1 Transmitter

The generation of entangled photons is a key element in the radar transmitter. To this end, a non-linear crystal is used to generate entangled photons. In other words, it splits an incoming photon into two entangled photons. In this view, semiconductor nanostructures are implemented to create entangled photons. Particularly, the deterioration of bi-excitonic situations in quantum dots contain, *inter-band transitions* between valence and conduction bands. In this circumstance, generate entangled photons can be made with frequencies in the microwave spectra. Consequently, microwave photons are created from impulsive descending conversions between single-particle levels in a quantum dot [67,70,80].

Entangled photons generation can be achieved by utilizing an array of quantum dots, which are connected to dual electron reservoirs and injected in a cylindrical microwave resonator. To simplify this concept, let assume four quantum dots (Fig. 5.11), which are labeled as d_1^O, d_2^O, d_3^O and d_4^O. In this view, d_2^O, and d_3^O are used to deliver distinctive initial and final states for an electron in the conduction band. However, d_1^O and d_4^O afford dual decay paths [81].

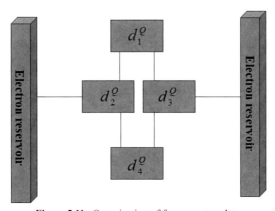

Figure 5.11. Organization of four quantum dots.

At the quantum dot d_2^Q, the electron begins. However, electron tunnels into excited states of quantum dots d_1^Q and d_4^Q, respectively. In this circumstance, the symmetric occurs and is formulated as:

$$|\Psi\rangle = \frac{1}{\sqrt{2}}\left(\left|d_1^Q\right\rangle + \left|d_4^Q\right\rangle\right) \tag{5.10}$$

Specifically, the electron can be in the distracted state of quantum dots d_1^Q or d_4^Q, with identical probability. Subsequently, the electron position deteriorates into pulverized states of quantum dots d_1^Q or d_4^Q and in the procedure discharges dual photons. If the electron deteriorates from d_1^Q, then both photons have spiral circular polarization states. Alternatively, if the electron deteriorates from d_4^Q, then both photons have clockwise circular polarization states. Then, the quantum state of this phase can be defined as:

$$|\Psi\rangle = \frac{1}{\sqrt{2}}\left(\left|d_1^Q\right\rangle \otimes |++\rangle + \left|d_4^Q\right\rangle \otimes |--\rangle\right) \tag{5.11}$$

The electron states can be determined from:

$$|\Psi_\pm\rangle = \frac{1}{\sqrt{2}}\left(\left|d_1^Q\right\rangle \pm \left|d_4^Q\right\rangle\right) \tag{5.12}$$

Conversely, the photonic states can be determined from:

$$|\Phi_\pm\rangle = \frac{1}{\sqrt{2}}\left(|++\rangle \pm |--\rangle\right) \tag{5.13}$$

The combination of Equations 5.12 and 5.13 deliver the generation of quantum photons and electron which can be described as:

$$|\Psi\rangle = \frac{1}{\sqrt{2}}\left(|\psi_+\rangle|\Phi_+\rangle + |\psi_-\rangle|\Phi_-\rangle\right) \tag{5.14}$$

The coupling with the dot d_3^Q performs to separate the photons from the electron conditions. Moreover, the antisymmetric state $|\Phi_+\rangle$ is restrained owing to disparaging interference. Subsequently, the state of the photons is located by acquiring the photon state's ledge $|\Psi\rangle$ with $|\psi_+\rangle$. In this circumstance, the radiated or transmitted photon is formulated as:

$$(\psi_+|\Psi\rangle\alpha|\Phi_+\rangle) = \frac{1}{\sqrt{2}}\left(|++\rangle + |--\rangle\right) \tag{5.15}$$

Equation 5.15 demonstrates that the radiated photons occur with utmost entanglement.

5.11.2 Receiver

The receiver is made up of novel metamaterial which can absorb the reflected microwave photons and can be used to perform single-shot photodetection. A novel metamaterial is critical engineering materials with specific physical characteristics

[82]. For instance, the metamaterial lens, located in metamaterial antenna systems, which is depleted as an effective coupler to the exterior, radiated microwave photons. In this regard, it is focused on radiation alone or from a microstrip transmission line into transmitting and receiving components. Subsequently, it can be exploited as a response device [21,67,70,80].

Conversely, the receiver is able to acquire or gain a microwave photon from the initial state $|0\rangle$ and transmitted to the new state $|1\rangle$. Consequently, this mechanism deteriorates into an enduring steady ground state $|g\rangle$ with an attenuation rate Γ. In this regard, the microwave photon signifys that the swellings come down to a consideration extent of those absorbers existing in the ground state (Fig. 5.12). Consequently, this sensor acts as a photographic film in the sense that once a photon has been absorbed by the metamaterial, the detector is turned into a steady and mesoscopically distinct state. In this view, a small operational bandwidth can be correlated to the attenuation rate Γ as:

$$\Delta\omega < \Gamma^{-1} \tag{5.16}$$

Equation 5.16 reveals the circumstance of that bandwidth $\Delta\omega$ be small in comparison to the time it takes to receive a reflected microwave photon. In this

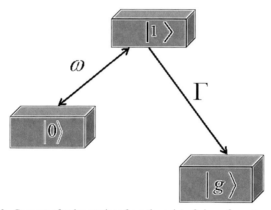

Figure 5.12. Concept of radar receiver from the point of view of quantum mechanics.

circumstance, the long wavepackets can be affected due to the occurrence of decay rate with a few frequencies.

5.12 Quantum Radar Illumination

Quantum photon entanglement when interacting with targets create reflected microwave photons that are known as quantum photon illumination which can preserve the information of any objects. In spite of the original entanglement being completely destroyed by a lossy and noisy environment. In other words, the interaction between the signal photons and the target can be modeled as a beam splitter with small reflectivity.

In this sense, noise is vaccinated into the radar system with an average photon number N_p into each frequency and polarization. For instance, thermal added radiation from the electric circuit in the antenna can be a source of noise, i.e., thermal noise [82].

Let us assume that the bandwidth $\Delta \omega$ and D_T a temporal detection window, in practice the sensor can distinguish between different radiation modes M as:

$$M \approx \Delta \omega \times D_T \tag{5.17}$$

The photodetector can observe one noise photon per detection event under the circumstance of;

$$M \times N_p \ll 1 \tag{5.18}$$

$N_p \ll 1$ corresponds to the state where the thermal emission is significantly beneath the signal photon energies. What are the impacts of non-entangled and entangled signal microwave photons signals on imagining oil spill? The answer to this question will be addressed in length in the following chapters.

Generally, the fundamental structure of quantum illumination is target detection. In this context, the radar antenna, for example, organizes dual entangled systems, called the signal and idler. The idler is recollected while the signal is radiated to probe the existence of a low-reflectivity object in a region with a bright background noise. The reflection from the object is then joined with the recollected idler system in a joint quantum measurement, which provides binary possible consequences: target existing or target absent. More specifically, the investigative process recurs numerous times so that countless sets of signal-idler systems are composed at the receiver for joint quantum detection.

Theories of Synthetic Aperture Radar

6.1 What is Meant by Aperture?

The keyword "aperture" signifies the diameter of the lens **exposure** opening. The aperture of the camera then determines the area through which light is collected [83]. Aperture, consequently, is restrained in focal length (f) stops. In this understanding, the smaller aperture receives less reflected light (Fig. 6.1). In other words, the blurred image develops in an extensively exposed aperture (Fig. 6.1). On the contrary, the sharp dark image are produced on smaller aperture because of an excessive depth of field.

The quantity of light netted by a lens is proportional to the extent of the aperture, equivalent to:

$$A = \pi \left[0.5d \right]^2 = \pi \left[\frac{f}{2n} \right]^2 \tag{6.1}$$

where A is the area, d is aperture diameter, f is the focal length and $n = fd^{-1}$. The focal length value is not essential when matching dual lenses of the same focal length; a rate of 1 can be depleted instead, and the other factors can throw down additionally, separation area proportion to the reciprocal square of the f-number n.

On the contrary, the minimal aperture does not rely on the focal length—it is restrained through how narrowly the aperture closes, not the lens layout—and is alternatively chosen based on practicality: very minor apertures have decreased

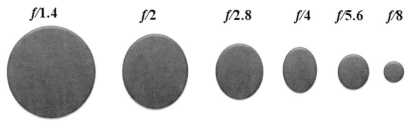

Figure 6.1. Picture quality based on the aperture.

sharpness owing to diffraction, whilst the presented depth of field is not usually useful, and consequently there is little gain in the use of such apertures. In view of that, Digital Single Lens Reflex "DSLR" lenses normally have a minimum aperture of $f/16$, $f/22$, or $f/32$, whilst giant structure can also go down to $f/64$, as reflected in the identity of Group $f/64$. The depth of field is a vast issue in macro photography, however, and there one sees smaller apertures.

6.2 Antenna Aperture

Consistent with the above perspective, the antenna aperture can be defined as an operational capacity or receiving cross-section, which is a quantity of how operative an antenna is at receiving the power of electromagnetic radiation, for instance, radio waves. In other words, the aperture is described as the area, oriented perpendicular to the path of an incoming electromagnetic wave, which would intercept the identical quantity of energy from that wave as it is formed by means of the antenna receiving it [84].

Let us consider a point x receives a beam of electromagnetic radiation, which has a power flux density or an irradiance $S_I(x)$. In this view, an irradiance is defined as the quantity of energy passing across a unit area of one square meter. The antenna aperture A_e can mathematically be formulated as:

$$A_e = \frac{P_0}{S_I(x)} \tag{6.2}$$

where P_0 is the power radiated by an antenna device when irradiated by undeviating field of power density $S_I(x)$ watts per square meter. Equation 6.2 reveals that the power received by an antenna (in watts) is equivalent to the power density of the electromagnetic energy (in watts per square meter), multiplied by its aperture (in square meters). The greater an antenna's aperture, the extra power it can gather from a particular electromagnetic field. To actually attain the expected power accessible to P_0 polarization of the incoming waves must match the polarization of the antenna, and the load (receiver) must be impendent rivaled to the antenna's feed point impedance. A_e can be implemented to determine the function of a transmitting antenna also. In this view, A_e is a function of the direction of the electromagnetic wave relative to the alignment of the antenna, since the gain of an antenna G fluctuates in relation to its beam patterns. One of the key parameters of an antenna gain is directivity. In this context, directivity is a constraint of an antenna or optical system, which quantifies the level to which the beam emitted is focused in a single direction. Moreover, it computes the energy density the antenna emits in the path of its strongest emission, as opposed to the energy density emitted by means of a perfect isotropic radiator (i.e., which emits homogeneously in entire directions) radiating the identical entire power [83–86].

It can be seen that the gain G is correspondingly equivalent to the ratio of

$$G = \frac{A_e}{A_{iso}} \tag{6.3}$$

where A_{iso} presents the aperture of a lossless isotropic antenna, which is $A_{iso} = \dfrac{\lambda^2}{4\pi}$.

In this regard, an isotropic radiator is a theoretical factor source of electromagnetic or radio waves, which emits the identical beams in all directions. It has no desired path of emission. It emits homogeneously in entire directions over a sphere centered on the cause. Isotropic radiators are implemented as orientation radiators with which different sources are equated, for instance, in identifying the gain of the antennas (Fig. 6.2). A coherent isotropic radiator of electromagnetic waves is theoretically incredible, however, incoherent radiators can be reconstructed [84–86].

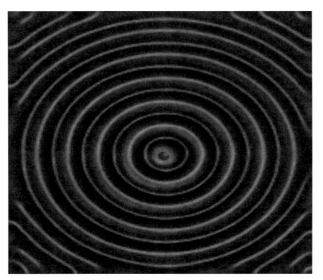

Figure 6.2. Isotropic radiation.

In this understanding, the gain antenna as a function $A_{iso} = \dfrac{\lambda^2}{4\pi}$ is expressed as:

$$G = 4\pi A_e \lambda^{-2} \tag{6.4}$$

where λ is the wavelength of radio waves. The antenna gain with a physical aperture of the area $A_{physical}$ is formulated as:

$$G = 4\pi A_{physical} e_\alpha \left[\lambda^{-2} \right] \tag{6.5}$$

where e_α is the antenna's *aperture efficiency* ($e_\alpha = \dfrac{A_e}{A_{physical}}$), in this sense, Equation 6.5 shows that the antennas with large effective apertures are high gain antennas with small angular beam widths. In other words, the receiving antennas, are furthermost precise to radio waves approaching from one direction and are considerably less precise to ones propagating from other directions. Since transmitting antennas, furthermost of their radiation is emitted in a fine beam in one direction, and tiny in other directions. Even though these terms can be depleted as a function of direction, when no direction is identified, the gain and aperture are assumed to denote the antenna's axis of maximum gain or the antenna boresight. In this context, the antenna boresight is the alignment of equilibrium of the parabolic dish, and the antenna beam pattern, i.e., the main lobe, which is symmetrical approximately of the boresight axis. In this regard, the majority of antennas boresight axis is permanent by their form and cannot be reformed (Fig. 6.3) [83–85].

There are certain sorts of antennas such as monopoles and dipoles, that are not defined by a physical area, which involves thin rod conductors. In this sense, the aperture tolerates no apparent relation to the dimension or area of the antenna. An alternative quantity of antenna gain that has a greater relationship to the physical structure of such antennas is *effective length* l_e calculated in meters, which is termed for a receiving antenna as:

$$l_e = V_0 E_s^{-1} \tag{6.5}$$

where V_0 presents the open circuit voltage acting across the antenna's terminals, and E_s presents the electric magnitude of the radio signal in volts per meter at the antenna. Equation 6.5 explains that the longer l_e, the more electrical power, which the antenna can receive. Consequently, an antenna's gain increase in relation to the square off l_e, which involves the antenna radiation resistance. Thus, this quantity is more of a hypothetical than practical rate and is not, by itself, a valuable digit of distinction concerning an antenna directivity [85–87]. In this understanding, in Synthetic Aperture Radar (SAR) systems a very long antenna aperture is synthesized resulting in fine along-track resolution. For a synthesized-aperture length, L, the along-track resolution, Δy, is

$$\Delta y = \lambda R \times [2L]^{-1} \tag{6.6}$$

For instance, ERS-1/2 has an antenna aperture length of 10 m with 1 m diameter. In fact, ERS-1/2 has a wavelength of 5.6 cm and a slant range of 845 km. In this regard, SAR uses an array of real antennas to synthesize the effect of a very long antenna. Conversely, the short antenna delivers a wide footprint, while a long antenna delivers a narrow footprint (Fig. 6.4).

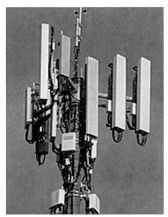

Figure 6.3. Example of antennas boresight.

Short antenna   **Long antenna**

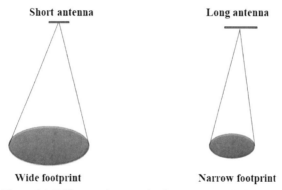

Wide footprint   **Narrow footprint**

Figure 6.4. Differences between the short antenna and a long one.

Generally, for real array imaging radar, its long antenna creates a fan beam illuminating the ground below. The along-track resolution is resolved by the beam width, while the across resolution is determined by the pulse length. The larger the antenna, the finer the detail that the radar can resolve.

6.3 Real and Synthetic Aperture Radar

Imaging radar is categorized further into Real Aperture Radar (RAR) and Synthetic Aperture Radar (SAR). In this view, RAR transmits a narrow-angle beam of the pulse radio wave in the range direction at right angles to the flight direction, which is known as the azimuth direction (Fig. 6.5). RAR receives the backscattering from the objects, which is transformed into a radar image from the received signals. Figure 6.5 reveals that the strip of terrain to be imaged is from point A to point B. Point A being nearest to the nadir point is said to lie at near range and point B, being the furthest, is said to lie at far range. The distance between A and B defines the swath width. The distance between any point within the swath and the radar is called its slant range. The ground range for any point within the swath is its distance from the nadir point (a point on the ground directly underneath the radar) [89].

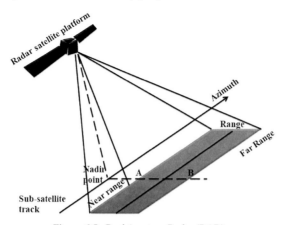

Figure 6.5. Real Aperture Radar (RAR).

   Usually, the reflected pulse will be arranged in the order of return time from the targets, which corresponds to the range direction scanning. The resolution in the range direction depends on the pulse width. However, if the pulse width is made small, in order to increase the resolution, the signal to noise ratio S/N of the return pulse will decrease because the transmitted power also becomes low. Therefore, the transmitted pulse is modulated to chirp with high power (Fig. 6.6), but wide-band, which is received through a matched filter, with the reverse function of transmission, to make the pulse width very narrow and high power. This is called pulse compression or de-chirping (Fig. 6.7). By making the pulse compression, with an increase of frequency in transmission, the amplitude becomes times bigger, and the pulse width becomes narrower. This method is called range compression.

   The resolution in the azimuth direction is identical to the multiplication of the beam width and the distance to a target. As the resolution of the azimuth direction increases with shorter wavelength and bigger antenna size, a shorter wavelength and a bigger antenna are used for higher azimuth resolution. However, as it is difficult to attach such a large antenna, as required for instance, a 1 km diameter antenna in order to obtain 25 meters resolution with L band, i.e. (= 25 cm), and 100 km distance from a target, a real aperture radar therefore, has a technical limitation for improving the azimuth resolution [90].

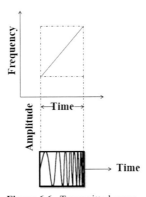

Figure 6.6. Transmitted wave.

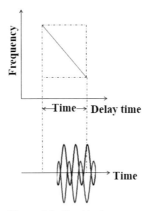

Figure 6.7. De-chirping wave.

Nevertheless, the image constructed by SLAR is deprived in azimuth resolution. For SLAR the smaller the azimuth beamwidth, the finer the azimuth resolution. With the intention of obtaining a high-resolution image, one has to resort either to an impractically long antenna or to employ wavelengths, so short that the radar must contend with severe attenuation in the atmosphere. In airborne application, particularly the antenna size and weight are restricted.

Similarly, the antenna length of radar partly specifies the area through which it collects radar signals. The length of the antenna is also called the aperture. In general, the larger the antenna, the more information you can obtain about a particular viewed object. With more facts, it can create a better image of that object (the improved resolution). It is ridiculously expensive to place very large radar antennas in space. To overcome this problem, a SAR used the synthetic aperture antenna. This means that the short antenna with their attended wide beam can be made to react as though they are very long. In fact, the SAR is able to transmit several hundred pulses while its parent spacecraft passes over a particular object [91]. In other words, another way of achieving better resolution from the radar is signal processing. Synthetic Aperture Radar (SAR) is a technique which uses signal processing to improve the resolution beyond the limitation of physical antenna aperture. In SAR, the forward motion of an actual antenna is used to 'synthesize' a very long antenna. SAR allows the possibility of using longer wavelengths and still achieving good resolution with antenna structures of reasonable size.

Compared to real aperture radar, Synthetic Aperture Radar (SAR) synthetically increases the antenna's size or aperture to increase the azimuth resolution, though the same pulse compression technique is adopted for range direction. Synthetic aperture processing is a complicated data process of receiving signals and phases of moving targets with a small antenna [92]. Figure 6.8 illustrates the geometry of the real aperture radar. The strip of terrain to be imaged is from point A to point B. Point A being nearest to the nadir point is said to lie at near range and point B, being furthest, is said to lie at a far range.

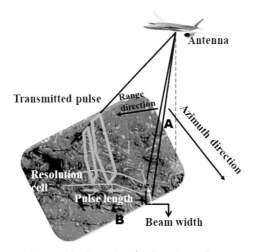

Figure 6.8. Geometry of real aperture radar.

In addition, the distance between A and B defines the swath width. The distance between any point within the swath and the radar is called its slant range. The ground range for any point within the swath is its distance from the nadir point (the point on the ground directly underneath the radar) (Fig. 6.8).

6.4 Radar Resolution

6.4.1 Spatial Resolution

Spatial resolution is the keystone to determine the physical object characteristics imagined in radar sensors. In this context, it is known as the potential of the radar sensor to identify the close range between two objects as separate points. For instance, if a certain radar system is capable of discriminating two closely spaced internal wave crests as separate, a lower resolution system may only imagine one internal wave crest. The internal wave is extremely clear with the fine lower resolution of 25 m as compared to 30 m and 100 m resolution, respectively (Fig. 6.9).

SAR spatial resolution is computed in the azimuth and range directions. Furthermore, it is defined by the characteristics of the radar system and sensor. Resolution in the azimuth direction is, theoretically, one-half the length of the radar antenna. In the context of signal processing, azimuth resolution is independent of range. Range resolution, consequently, is governed by the frequency bandwidth of the transmitted pulse and thus by the time duration (width) of the range-focused pulse. For instance, large bandwidths yield a small focused pulse width [93]. Angular resolution is the minimum angular separation at which two equal targets at the same range can be separated (Fig. 6.10). The angular resolution as a distance between two objects relies on the slant-range and is formulated as [94]:

$$S_A \leq 2R \cdot \sin\frac{\theta}{2} \ [m] \tag{6.7}$$

θ being antenna beam width (Theta), R is slant range aims-antenna, and S_A is the angular resolution as the distance between the two targets. The angular resolution characteristics of radar are determined by the antenna beamwidth represented by the -3 dB angle θ which is defined by the half-power (-3 dB) points [95]. The half-power points of the

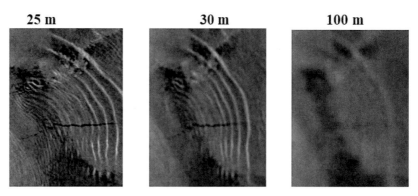

Figure 6.9. Different SAR image resolutions.

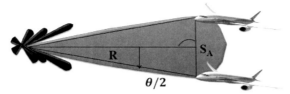

Figure 6.10. Angular resolution.

antenna radiation pattern (i.e., -3 dB beamwidth) are normally specified as the limits of the antenna beam width for the purpose of defining angular resolution; two identical targets at the same distance are, therefore, resolved in angle if they are separated by more than the antenna beam width. In this regard, the smaller the beam width, θ, the higher the directivity of the radar antenna, and the better the bearing resolution [96].

6.4.2 Slant and Ground Range Resolution

As the range resolution of imaging SARs depends on signal pulse length, the actual distance resolved is the distance between the leading and the trailing edge of the pulse. These pulses can be shown in the form of signal wavefronts propagating from the SAR sensor. When these wavefront arcs are projected to intersect a "flat" Earth's surface, the resolution distance in the ground range is always larger than the slant range resolution (Fig. 6.11). The ground range resolution increases substantially at small incidence angles [97].

Therefore, the theoretical range resolution of a radar system is estimated via:

$$S_r = \frac{c_0 \cdot \tau}{2} \; [m] \tag{6.8}$$

where, c_0 is the speed of light, τ is the transmitter pulse width, and S_r is the range resolution as a distance between the two targets. In a pulse compression system, the range-resolution of the radar is given by the bandwidth of the transmitted pulse (B_{tx}), not by its pulse width [97],

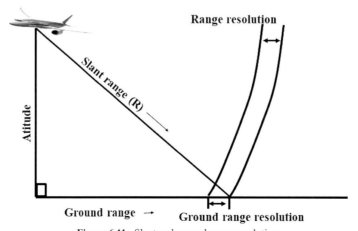

Figure 6.11. Slant and ground range resolution.

$$S_r \geq \frac{c_0}{2B_{tx}} \; [m] \qquad\qquad (6.9)$$

where S_r is the range resolution as the distance between the two targets, and B_{tx} is the bandwidth of the transmitted pulse. This allows an identical high resolution to be obtained with a long pulse, thus with a higher average power [93–99].

6.4.3 Resolution Cell

Consistent with Moreira [92], the range and angular resolutions lead to the resolution cell. The meaning of this cell is very clear: unless one can rely on eventual different Doppler shifts, it is impossible to distinguish two targets which are located inside the same resolution cell. The shorter the pulse with τ (or the broader the spectrum of the transmitted pulse) and the narrower the aperture angle are, the smaller the resolution cell, and the higher the interference immunity of the radar station is.

Ground-range resolution is the weakest in the near-range portion of the SAR image and the best in the far-range sector. The depression angle β brightens the near-range portion of the swath, although, small depression angles illuminate the far-range sector of thebeam. The relationship between ground resolution and depression angle is casted as:

$$R_{GR} = \frac{\tau c}{2\cos \beta} \qquad\qquad (6.10)$$

where R_{GR} is the ground resolution, c is the speed of light and β is the depression angle. Equation 8.4 indicates that the pulse duration, ground range and beam width are ruled by the size of the ground resolution cell. Pulse duration and ground range, consequently, prescribe the spatial resolution in the path of energy transmission which denotes the range resolution. The spatial resolution in the direction of flight is known as azimuth resolution and is determined by beamwidth [94–100].

6.4.4 Ambiguous Range

The main difficulty with range measurement and pulsed radars is how to unambiguously determine the range to the target when the target returns a strong echo. An ambiguous range results from transmission of a sequence of pulses. In this view, Pulse Repetition Interval (PRI) is created and is known as the spacing between transmit pulses. It is also described as a Pulse Repetition Frequency (PRF). Consequently, the delay time is created because of a spacing between transmit and return pulses. In this view, the delay time is generated and causes an ambiguity along the range direction or uncertainty. The unambiguous range R_A as a function of delay time τ_{PRI} is defined as:

$$R_{amb} = \frac{c\tau_{PRI}}{2} \qquad\qquad (6.11)$$

In Equation 6.11, the radar can determine its range unambiguously when the object range is smaller than R_{amb}. On the contrary, an impossible range of unambiguously accounts when the object range is larger than R_{amb}. Under these conditions, the radar must have PRI to be greater from the range delay frequented with the longest target ranges to circumvent range ambiguities. An alternative method to avoid the ambiguous range problem is based on multiple PRIs with waveforms. In this case, the waveforms assist to change the spacing between transmit pulses and detect that the target range is ambiguous. Then it can be easy to ignore the return pulse. The advanced method to resolve range delays, is based on range resolve algorithm to calculate the precise target range [93,100].

6.5 Range-Rate Measurement (Doppler)

Doppler frequency can measure the range rate. In fact, the Doppler frequency arises from the frequency differences by the spacing between the transmitted and received signals. Consider the aircraft linear velocity of v, with different time of dt and varied range of dR as a function of plane movements from A to B (Fig. 6.12). In this sense, the range rate is determined from:

$$\dot{R} = \frac{dR}{dt} \tag{6.12}$$

The transmitted pulse can mathematically be written as:

$$v_T(t) = \text{rect}\left[\frac{t}{\tau_p}\right]\cos(2\pi f_c t) \tag{6.13}$$

where

$$\text{rect}\left[\frac{t}{\tau_p}\right] = \begin{cases} 1 & 0 \le t < \tau_p \\ 0 & \text{elsewhere} \end{cases}. \tag{6.14}$$

being, f_c is the *carrier frequency* of the radar, $\tau_p = 1\ \mu s$ is pulse width, and t is a time of the transmitted pulse. On the contrary, the received pulse as a function of delay time or range rate v_R is defined as:

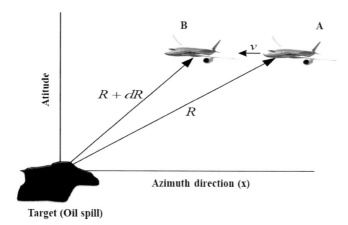

Figure 6.12. Range rate measurement by Doppler.

$$v_R(t) = \xi v_T(t - \tau_R) \tag{6.15}$$

where ξ is the amplitude scaling factor, and τ_R, is the range delay which is given by

$$\tau_R(t) = \frac{2R(t)}{c} \tag{6.16}$$

Then the Doppler frequency is a function of a range delay and can mathematically be given by

$$v_R(t) = \xi \, \text{rect}\left[\frac{t - \tau_R}{\tau_p}\right] \cos\left(2\pi(f_c + f_d)t + \phi_R\right) \tag{6.17}$$

where $\phi_R = -2\pi f_c(2R/c)$ is the phase shift owing to a range delay and $f_d = -f_c(2\dot{R}/c)$ is the *Doppler frequency* of the target which can be rewritten as a function of radar wavelength λ,

$$f_d = -2\dot{R}/\lambda. \tag{6.17.1}$$

Equation 6.17 demonstrates that the frequency of the backscatter signal is $f_c + f_d$ in preference to simply f_c. The Doppler frequency, f_d, therefore, can be computed based on the comparison between the frequency of the transmit signal to the frequency of the received signal. Once we have f_d, we can compute the range-rate from Equation 6.17.1.

In practice, computing Doppler frequency is not as simple as indicated above. The difficult deceptions in the relative magnitudes of f_d and f_c. To understand this, let us assume a target that is traveling at approximately 300 m/s. In this case, the target is flying nonstop toward the radar with X-band, i.e., its specific carrier frequency of $f_c = 10$ GHz, with the intention of $\dot{R} = -v = -300$ m/s. Based on the speed of light, the wavelength of X-band is given by:

$$\lambda = \frac{c}{f_c} = \frac{3 \times 10^8}{10 \times 10^9} = 0.03 \text{ m} \tag{6.18}$$

and

$$f_d = -\frac{2\dot{R}}{\lambda} = -\frac{2(-300)}{0.03} = 20^4 \text{ Hz} = 20 \text{ KHz.} \tag{6.19}$$

Both Equations 6.18 and 6.19 show that f_d is much smaller than f_c.

Hypothetically, computing Doppler frequency is not as easy as implied in Equation 6.17. The difficulty is because of the comparative magnitudes of f_d and f_c. Though the quantity of Doppler frequency is difficult, it is feasible. To compute the Doppler frequency, the transmit signal must be extremely long (in the order of ms rather than μs) or the Doppler frequency must rely on the processing of numerous signals.

6.6 Ambiguity Function of SAR

starthe ambiguity function, which is symbolized as $|\chi(\tau, f)|$, is principally exploited to extend a perception of how a signal processor reacts to a specified backscatter signal. As specified in the symbolization, the independent parameters of the ambiguity function are time (τ) and frequency (f). In this regard, τ is generally allied with object range and f is typically connected to target Doppler. In fact, the code comprises a magnitude symbol, i.e., $|\ |$, which is implemented to specify that the amplitude of the signal processor output is characterized. In this circumstance, the signal processor must be matched to the transmitted waveform. In this understanding, when the signal processor is not matched to the transmitted waveform, the ambiguity function context has to denote to the cross ambiguity function [100].

Let us assume that the transmitted waveform be represented by the base-band signal, $u(t)$. In this view, the normalized (base-band) signal received from a (constant range-rate) target at a range R and range-rate $\dot{R}$ is formulated as:

$$v_R(t) = u(t - \tau_R) e^{j2\pi f_d t} \tag{6.20}$$

where $\tau_R = 2R/c$ is the range delay and $f_d = -2\dot{R}/\lambda$ is the target Doppler frequency. λ is the wavelength of the transmitted signal and c is the speed of light as mentioned in Equation 6.17. Remember that if a real signal is $v_{RF}(t) = A(t)\cos(\omega_{RF} t - \phi(t))$, then the base-band depiction of that signal is $v_B(t) = A(t)e^{j\phi(t)}$. Specifically, the base-band signal confines the amplitude and phase modulation of the definite signal in the formula of a complex conspiracy. Therefore, base-band symbolization is correspondingly designated as complex signal notation. In this context, Fig. 6.13 reveals the signal processor configuration, which is implemented in deriving the ambiguity function. In this sense, $h(t)$ presents a low-pass function and f_s represents the theory as the frequency to which the signal processor is tweaked. Accordingly, the inclusive signal processor is a band-pass filter centered at f_s. This is specified in the frequency response of $h(t)$ and the frequency response of the signal processor (Fig. 6.14). Let us consider $h(t)$ in the conditions of the waveform to which it is matched. Under this assumption, $h(t)$ is formulated as:

$$h(t) = Kv^*(t_0 - t) \tag{6.21}$$

and set $K = 1$ and $t_0 = 0$ to obtain

$$h(t) = v^*(-t). \tag{6.22}$$

here $v(t)$ presents the waveform to which the signal processor is matched. In this circumstance, the impulse response of the signal processor is computed as:

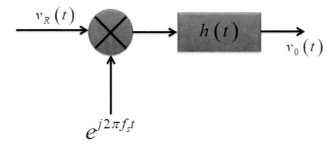

Figure 6.13. Signal Processor.

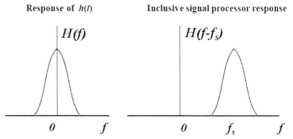

Figure 6.14. The frequency response of the signal processor.

$$v_o(\gamma) = \int_{-\infty}^{\infty} u(t)v^*(t + \tau_R - \gamma)e^{j2\pi(f_d - f_s)(t+\tau_R)}dt$$

$$= e^{j2\pi f_s \tau}e^{j2\pi f_d \tau_R}\int_{-\infty}^{\infty} u(t)v^*(t+\tau)e^{j2\pi ft}dt$$

(6.23)

Equation 6.23 reveals that the parameters τ and f are frequently designated the mismatched range and Doppler of the ambiguity function. More precisely, τ signifies the modification between the target range delay and the time at which the signal processor output is considered. When $\tau = 0$, the matched range is achieved. In other words, the signal processor at a time equal to the time delay of the target. Therefore, f signifies the variance between the target Doppler frequency and the frequency to which the signal processor is matched. In this circumstance, the signal processor is matched to the target Doppler, or vice-versa if $f = 0$. In this sense, the matched Doppler is obtained [101].

Let us change the dependent variable from $v_o(\gamma)$ to $\chi(\tau, f)$, then the following formula can be given as:

$$\chi(\tau, f) = e^{j2\pi f_s \tau}e^{j2\pi f_d \tau_R}\int_{-\infty}^{\infty} u(t)v^*(t+\tau)e^{j2\pi ft}dt.$$

(6.24)

Finally, the ambiguity function is obtained as the absolute value of Equation 6.24:

$$|\chi(\tau, f)| = \left|\int_{-\infty}^{\infty} u(t)v^*(t+\tau)e^{j2\pi ft}dt\right|.$$

(6.25)

We frequently use singular terms with $|\chi(\tau, f)|$ for specific values of τ and f. In particular:

- If $f = 0$ to yield $|\chi(\tau, 0)|$, the matched Doppler, range cut of the ambiguity function. This is what we normally think of as the output of the classical matched filter.
- If $\tau = 0$ to yield $|\chi(0, f)|$, the matched range, Doppler cut of the ambiguity function.
- If $f = f_k$ to yield $|\chi(\tau, f_k)|$, a range cut at some mismatched Doppler of f_k.
- If $\tau = \tau_k$ to yield $|\chi(\tau_k, f)|$ a Doppler cut at some mismatched range of τ_k [100–102].

Generally, the ambiguity function delivers a prosperity of information about SAR waveforms and how they interact with the environment and the radar signal processor. Conversely, the equation for the ambiguity function of an unmodulated pulse of width *T* can be given as:

$$\left|\chi(\tau,f)\right| = \left|\int_{-\infty}^{\infty} \text{rect}\left[\frac{t-T/2}{T}\right]\text{rect}\left[\frac{t+\tau-T/2}{T}\right]e^{j2\pi ft}dt\right| \tag{6.26}$$

where $\text{rect}[x] = \begin{cases} 1 & x \le \frac{1}{2} \\ 0 & x > \frac{1}{2} \end{cases}$. $\tag{6.26.1}$

For $\tau < 0$ the rect functions overlap from $-\tau$ to T. Thus the ambiguity function becomes

$$\left|\chi(\tau,f)\right| = \left|\int_{-\tau}^{T} e^{j2\pi ft}dt\right| = \left|\frac{e^{j2\pi fT} - e^{-j2\pi f\tau}}{j2\pi f}\right|. \tag{6.27}$$

Let us factor $e^{j\pi f(T-\tau)}$ out of both terms on the far right side the result can be formulated as:

$$\left|\chi(\tau,f)\right| = \left|e^{j\pi f(T-\tau)}\left(\frac{e^{j\pi f(T+\tau)} - e^{-j\pi f(T+\tau)}}{j2\pi f}\right)\right| = \left|T+\tau\right|\left|\text{sinc}\left(f(T+\tau)\right)\right|. \tag{6.28}$$

Finally, since $\tau < 0$, $\left|\tau\right| = -\tau$ Equation 6.28 can be written as:

$$\left|\chi(\tau,f)\right| = \left|T-\left|\tau\right|\right|\left|\text{sinc}\left(f(T-\left|\tau\right|)\right)\right|. \tag{6.29}$$

Converesely, if $\tau \ge 0$, the rect functions overlap from 0 to $T-\tau$, thus the ambiguity function becomes

$$\left|\chi(\tau,f)\right| = \left|\int_{0}^{T-\tau} e^{j2\pi ft}dt\right| = \left|\frac{e^{j2\pi f(T-\tau)} - 1}{j2\pi f}\right|. \tag{6.30}$$

Let us refactor $e^{j\pi f(T-\tau)}$ out of both terms on the far right side, then Equation 6.30 can be written as [101]:

$$\left|\chi(\tau,f)\right| = \left|e^{j\pi f(T-\tau)}\left(\frac{e^{j\pi f(T-\tau)} - e^{-j\pi f(T-\tau)}}{j2\pi f}\right)\right| = \left|T-\tau\right|\left|\text{sinc}\left(f(T-\tau)\right)\right|. \tag{6.31}$$

Finally, since $\tau \ge 0$, $\left|\tau\right| = \tau$ Equation 6.31 can be written as:

$$\left|\chi(\tau,f)\right| = \left|T-\left|\tau\right|\right|\left|\text{sinc}\left(f(T-\left|\tau\right|)\right)\right| \tag{6.32}$$

Equation 6.32, which is the same result we obtained for $\tau < 0$. Let us compile all of the Equations 6.28 to 6.32, so the following mathematical expression is given by:

$$\left|\chi(\tau,f)\right| = \left(T-\left|\tau\right|\right)\left|\text{sinc}\left(f(T-\left|\tau\right|)\right)\right|\text{rect}\left[\frac{\tau}{2T}\right]. \tag{6.33}$$

Obviously, the matched Doppler and range cut, which are obtained by setting $f = 0$, is the similar form of the matched filter output. Specifically,

$$\left|\chi(\tau,0)\right| = \left(T - |\tau|\right)\text{rect}\left[\frac{\tau}{2T}\right]. \tag{6.34}$$

In this regard, the ambiguity function $\chi(\tau,f)$ can be interpreted as the correlation of $u(\tau)e^{j2\pi f\tau}$ with $v(\tau)$, which is formulated as:

$$\chi(\tau,f) = \int_{-\infty}^{\infty} u(t)v^*(t+\tau)e^{j2\pi ft}dt = \left(u(\tau)e^{j2\pi f\tau}\right) \otimes v(\tau) \tag{6.35}$$

The ambiguity function can be determined as a sequence of Doppler cuts. In this circumstance, note that, for some $\tau = \tau_k$, we can interpret $\chi(\tau_k,f)$ as the inverse Fourier transform of $u(t)v^*(t+\tau_k)$. The aggregation of these is the ambiguity function. In terms of Fourier transform, Equation 6.35 can be described as:

$$\Im\{\chi(\tau,f)\} = \Im\{\left(u(\tau)e^{j2\pi f\tau}\right)\} \Im^*\{v(\tau)\} \tag{6.36}$$

In this circumstance $u(t)v^*(t+\tau_k)$ is computed for various $\tau = \tau_k$ and then use the IFFT to approximate the appropriate inverse Fourier transform. Each of these is a Doppler cut at the mismatch range τ_k [102].

6.7 SAR Pulse Compression Waveforms

Pulse compression waveforms are a sort of waveform, which is associated with the signal processors to deliver excellent range resolution. In fact, the overall signal duration is long, which leads the output of the much shorter signal (Fig. 6.15). Thus, it tolerates SAR to accomplish virtuous SNR, at reasonable power levels and good range resolution. All pulse compression waveforms are dominated by constant amplitude, phase (or frequency) modulated waveforms. The mathematical description of these waveforms is formulated as:

$$u(t) = e^{j\phi(t)}\text{rect}\left[\frac{t}{\tau_p}\right]. \tag{6.37}$$

Equation 6.37 presents signal processor (or matched filter), which is matched to $u(t)$, by means of the terms of the ambiguity function, $v(t) = u(t)$. In this view, the pulse compression waveform is considered is the Low-Frequency-Modulation (LFM) waveform (Fig. 6.16). A quadratic phase function $\phi(t)$ is involved with this waveform and is given by:

$$\phi(t) = \pi bt^2. \tag{6.38}$$

Moreover, the derivative of $\phi(t)$ delivers the term linear frequency modulation, which is computed as:

$$\omega(t) = 2\pi f(t) = \frac{d\phi(t)}{dt} = 2\pi bt \tag{6.39}$$

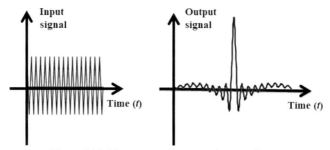

Figure 6.15. The concept of compression waveforms.

$$\omega(t) = 2\pi f(t) = \frac{d\phi(t)}{dt} = 2\pi bt \tag{6.39}$$

or

$$f(t) = bt, \tag{6.40}$$

In other words, the frequency $u(t)$ varies linearly with time. Besides, the total frequency excursion is $B = b\tau_p$. This indicates that $u(t)$ encloses frequencies, which extend over a band of $B = b\tau_p$. Accordingly, the LFM bandwidth is $B = b\tau_p$.

The linear frequency modulation or chirp, waveform has a rectangular amplitude modulation with pulse width T and linear frequency modulation with a swept bandwidth B applied over the pulse. The time-bandwidth product of the LFM waveform is equal to TB, where TB is the product of pulse width and swept bandwidth. The 3-dB width of the compressed pulse at the output of the matched filter is $\tau_3 = 0.886/B$ for large values of the time-bandwidth product. The peak time sidelobe level of the compressed pulse is -13.2 dB. In practice, for a waveform to modify as a pulse compression waveform, its range cut must have an equally narrow main lobe and equally low sidelobes. When a linear-FM waveform is desired, the phase samples follow a quadratic pattern and can be generated by two cascaded digital integrators. The input digital command to the first integrator defines this quadratic phase function. The digital command to the

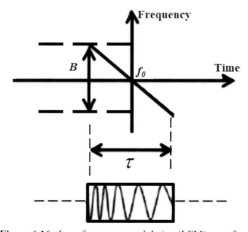

Figure 6.16. Low-frequency-modulation (LFM) waveform.

second integrator is the output of the first integrator plus the desired carrier frequency. This carrier may be defined by the initial value of the first integrator. The desired initial phase of the waveform is the initial value of the second integrator or else may be added to the second-integrator output.

The main lobe of the compressed pulse at the output of the matched filter has time or range, side lobes that occur within time intervals of duration τ_p, before and after the peak of the peak of the compressed pulse (Fig. 6.17). The time sidelobes can conceal targets, which would otherwise be resolved using a narrow uncoded pulse. In some cases, such as phase-coded waveforms or nonlinear frequency modulation waveforms, matched filter processing alone achieves acceptable time sidelobe levels. However, for the case of a linear frequency modulation waveform, the matched filter is generally followed by a weighting filter to provide a reduction in time sidelobe levels. In this case, the weighting filter results in a signal-to-noise ratio loss compared to that of matched filter processing alone.

Conversely, SAR systems are no longer limited to the LFM waveform; instead, SAR system capabilities can be extended to take advantage of the more complex processing associated with the nonlinear FM waveform. In this view, the nonlinear-FM waveform (Fig. 6.18) has several distinct advantages over LFM. It requires no frequency domain weighting for time sidelobe reduction because the FM modulation of the waveform is designed to provide the desired spectrum shape that yields the required time sidelobe level. This shaping is accomplished by increasing the rate of change of frequency modulation near the ends of the pulse and decreasing it near the center. This serves to taper the waveform spectrum so that the matched filter response has reduced time sidelobes. Thus, the loss in signal-to-noise ratio associated with frequency domain weighting (as for the LFM waveform) is eliminated.

Pulse compression waveforms consists of a concatenation of short pulses which is termed subpulses, where the phase, $\phi(t)$, is constant over the duration of each subpulse but fluctuates from subpulse-to-subpulse (Fig. 6.19). A specific of these *phase coded* waveforms is that the width of the main lobe of the range cut is double the width of the subpulses. The complete extent of the side lobes is twice the entire

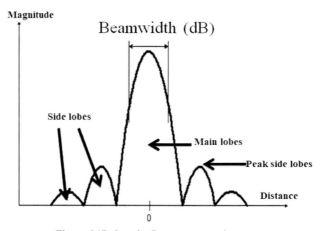

Figure 6.17. Impulse Response to a point target.

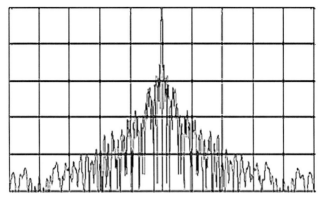

Figure 6.18. Nonlinear–frequency–modulation waveform.

length of the complete pulse, which equals to the width of the subpulses times the number of subpulses.

The mathematical description of the waveform can be written as:

$$u(t) = \sum_{k=0}^{N-1} e^{j\phi_k} \, \text{rect}\left[\frac{t - k\tau_{sp}}{\tau_{sp}}\right] \tag{6.41}$$

where the ϕ_k are the phases of the individual subpulses, τ_{sp} is the width of the individual subpulses and N is the number of subpulses that make up the overall pulse. The overall pulse width τ_p is $\tau_p = N\tau_{sp}$. In this understanding, the phase code on each subpulse is resolute by drawing random numbers. For instance, on each subpulse, one would draw a number from a random distribution that extends from -1 to 1. If the random number is positive, the subpulse would have a phase of 0 and if the random number is negative, the subpulse would have a phase of π. However, one cannot define in advance whether the resulting waveform will have sidelobes (in range-Doppler space) that are uniformly low. The only way to realize is to generate a waveform and form the ambiguity function. As noted previously, a common feature of the ambiguity function is that the volume of the ambiguity function is the same for all pulses of the same length, which are assumed constant amplitude pulses. Its peak value is also matching. In other words, if one decreases sidelobes in one area of range-Doppler space, the sidelobes must increase in other regions. With LFM pulses, low range-Doppler sidelobes can be obtained by "pushing" the volume along the ridge [101–105].

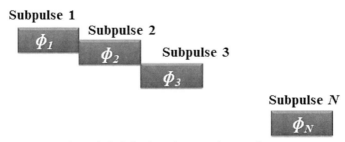

Figure 6.19. Subpulses of compression waveforms.

6.8 Range Compression

Let us consider an object travels along its path, the SAR transmits a linear-FM, or "chirp", pulse, which is defined as:

$$s(t) = \exp[\,j2\pi\,(f_c t + Kt^2/2)], \qquad |t| < \tau_p/2 \tag{6.42}$$

where, frequency (time derivative of phase) $f = f_c + K_t$, which is a linear function of time over the pulse duration τ_p with chirp rate K. This pulse contains:

Plane wave: $h(t) = e^{\,j2\pi ft}$ $\qquad\qquad\qquad\qquad\qquad\qquad\qquad$ (6.43)

Phase: $\phi = 2\pi ft$ $\qquad\qquad\qquad\qquad\qquad\qquad\qquad\qquad\qquad$ (6.44)

Frequency: $f = \dfrac{\dot{\phi}}{2\pi}$ $\qquad\qquad\qquad\qquad\qquad\qquad\qquad\qquad$ (6.45)

The chirp bandwidth (in range direction) is $B_R = K\tau_p$ since the frequency starts at $f_c - K\tau_p/2$, sweeps through all intermediate frequencies, and ends at $f_c + K\tau_p/2$. The scientific explanation of multi-pulse real transmitted signal can mathematically be written as:

$$p(t) = \sum_n s(t - nT_p), \tag{6.46}$$

where T_p presents the pulse repetition period and the sum contains mainly pulses for which the object is in the SAR beam. At a random time t, the SAR is at an approximately slant range $R(t)$ from a target point with image coordinates (x_c, R_c) at the SAR beam center [101,105]. The received pulse train is determined from:

$$v(t) = \sum_n a_n s[t - nT_p - 2R(t)/c],\ \left|t - nT_p - 2R(t)/c\right| \le \frac{\tau_p}{2}. \tag{6.47}$$

Consistent with Equation 6.47, the range compression is to correlate the received pulse with $s^*(t)$ (equivalently matched filtering with $s^*(-t)$) as:

$$g(t) = \int_{-\infty}^{\infty} s^*(t'-t)v(t')dt' \tag{6.48}$$

$$= \sum_n a_n \int_{-\infty}^{\infty} s^*(t'-t)s\!\left(t' - nT_p - 2R(t')/c\right)\!dt'$$

Consistent with above perspective, the variation of $R(t)$ within a chirp pulse is very minor, thus it can be assumed that $R(t) = R_n$, the distance between the sensor and the target at a chirp center time (Fig. 6.20). In this regard, the slant range $R(t)$ can be assumed constant throughout the period of one pulse width. Therefore, this postulation enables the common two-dimensional compression problem to be de-coupled into a sequence of two one-dimensional compression operations. In this understanding, one in fast time (range) and the other in slow time (azimuth). Subsequently, the slow time quantifies coordinates orthogonal to the range. In this sense, this method is termed as the rectangular algorithm [99,101,103]. The center of the returned chirp signal from a point target is at t_n, which is defined as:

$$t_n = nT_p + 2R_n/c. \tag{6.49}$$

The matched filtering of Equation 6.49 is expressed as:

$$g(t) = \sum_n a_n \int_{-\infty}^{\infty} s^*(t'-t)s(t'-t_n)dt' \tag{6.50}$$

However, the real integration is not constrained between $(-\infty,\infty)$, for instance, there is a time limit in both $s(t)$ and $v(t)$. Consequently, the range of effective integration also relies on t. In this context, Fig. 6.21 can be used to obtain the range of integration.

Conversely, the constraints of the integration yields:

i) If $\left|t - t_n\right| < \tau_p$, then

$$g(t) = \sum_n a_n \int_{\frac{t+t_n}{2} - \frac{\tau_p - |t-t_n|}{2}}^{\frac{t+t_n}{2} + \frac{\tau_p - |t-t_n|}{2}} \exp\left\{-j2\pi\left[f_c(t'-t) + K(t'-t)^2/2\right]\right\}$$

$$\cdot \exp\left\{j2\pi\left[f_c(t'-t_n) + K(t'-t_n)^2/2\right]\right\}dt' \tag{6.51}$$

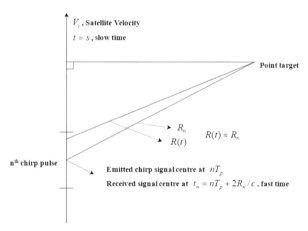

Figure 6.20. Concept of range compression.

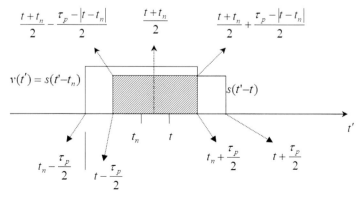

Figure 6.21. Computing range of integration.

ii) else $g(t) = 0$.   (6.52)

In these circumstances, Equation 6.51 can be evaluated as:

$$g(t) = \sum_n a_n \exp\left\{-j2\pi\left[-f_c t + Kt^2/2\right]\right\} \cdot \exp\left\{j2\pi\left[-f_c t_n + Kt_n^2/2\right]\right\}$$

$$\int_{\frac{t+t_n}{2}+\frac{\tau_p-|t-t_n|}{2}}^{\frac{t+t_n}{2}+\frac{\tau_p-|t-t_n|}{2}} \exp\left\{-j2\pi\left[f_c t' + K(t'^2 - 2tt')/2\right]\right\}\exp\left\{j2\pi\left[f_c t' + K(t'^2 - 2t_n t')/2\right]\right\}dt'$$

$$= \sum_n a_n \exp\left\{j2\pi\left[f_c(t-t_n) - K(t^2 - t_n^2)/2\right]\right\}\int_{\frac{t+t_n}{2}-\frac{\tau_p-|t-t_n|}{2}}^{\frac{t+t_n}{2}+\frac{\tau_p-|t-t_n|}{2}} \exp\left[j2\pi K(t-t_n)t'\right]dt'$$

                    (6.53)

Let us assume that $t' - \dfrac{t+t_n}{2} = t''$, then

$$g(t) = \sum_n a_n \exp\left\{j2\pi\left[f_c(t-t_n) - K(t^2 - t_n^2)/2\right]\right\}\int_{-\frac{\tau_p-|t-t_n|}{2}}^{\frac{\tau_p-|t-t_n|}{2}} \exp\left[j2\pi K(t-t_n)(t'' + \frac{t+t_n}{2})\right]dt''$$

$$= \sum_n a_n \exp\left[j2\pi f_c(t-t_n)\right]\int_{-\frac{\tau_p-|t-t_n|}{2}}^{\frac{\tau_p-|t-t_n|}{2}} \exp\left[j2\pi K(t-t_n)t''\right]dt''$$ (6.54)

The sinc function is obtained by integration as:

$$g(t) = \sum_n a_n \exp\left[j2\pi f_c(t-t_n)\right] \cdot \frac{\sin\left(\tau_p - |t-t_n|\right)\pi K(t-t_n)}{\pi K(t-t_n)}$$ (6.55)

Equation 6.55 holds if $|t-t_n| < \tau_p$ otherwise $g(t) = 0$. The signal near to the maximum power of the sinc function ($|t-t_n| \ll \tau_p$) can be formulated as:

$$g(t) = \sum_n a_n \exp\left[j2\pi f_c(t-t_n)\right] \cdot \frac{\sin \tau_p \pi K(t-t_n)}{\pi K(t-t_n)}$$ (6.56)

In Equation 6.56, the summation is dropped so that the range compressed result of nth pulse, $g_n(t)$ can be casted as:

$$g_n(t) = a_n \exp\left[j2\pi f_c(t-t_n)\right] \cdot \frac{\sin \tau_p \pi K(t-t_n)}{\pi K(t-t_n)}$$ (6.57)

Substitute the following parameters into Equation 6.57, $\omega = 2\pi f$ and, then

$$g_n(t) = a_n \exp(j\omega_c t)\exp(-j2\pi n T_p)\exp(-j4\pi R_n/\lambda) \cdot \frac{\sin \pi K\tau_p(t-t_n)}{\pi K(t-t_n)}$$ (6.58)

Let us assume the synchronization of the detailed pulse waveform, which is the SAR is coherent. In effect, the term $\exp(-j2\pi n T_p)$ can be ignored as it is identical for all pulses [103–105]. Then eEuation 6.58 becomes:

$$g_n(t) = a_n \exp(j\omega_c t)\exp(-j4\pi R_n/\lambda) \cdot \frac{\sin \pi K \tau_p(t-t_n)}{\pi K(t-t_n)}$$ (6.59)

The envelope of Equation 6.59 has a 3 dB width (half the maximum power), which is determined as:

$$\delta t = \frac{1}{K\tau_p} = \frac{1}{B_R}.$$ (6.60)

Equation 6.60 reveals the slant range time resolution of the SAR system, which corresponds to slant range spatial resolution (or, range resolution in short) as:

$$\delta R = c/2B_R,$$ (6.61)

where $B_R = K\tau_p$ is the range chirp bandwidth. Hence, a large chirp bandwidth offers enhanced range resolution for a SAR system. Note that the range resolution holds with positive K, 'up-chirp'. In case of 'down-chirp', K should be replaced with $|K|$ [102–106].

6.9 Azimuth Compression

Once range compression, the SAR data are converted into the azimuth frequency domain by using Fourier Transform (FFT). There are dual significant constraints for azimuth compression: (i) Doppler centroid frequency f_{Dc} and (ii) Doppler rate f_R, which can be gained from the SAR satellite state vectors. These include the satellite's position and velocity vectors. Occasionally, those principles prerequisite to be recomputed more precisely, which can be achieved by means of the data itself. In this context, the Doppler centroid frequency f_{Dc} is revealed from the SAR data through the clutter lock approach [106]. Therefore, the Doppler rate f_R can be formed by an autofocus method implementing the azimuth subaperture correlation procedure. Multiple Pulse Repetition Frequency (PRF) ambiguity of Doppler centroid, which is important for range migration, can be resolved by the range supporter correlation approach. In this regard, azimuth spectral filtering can be smeared on dual SAR signals forming an interferometric pair to compensate decorrelation from dissimilar Doppler centroids of the dual SAR images [107]. The signal is then azimuth compressed by means of the identically matched filtering as the circumstance of range compression. To this end,a Hamming filter can then be smeared to diminish the side lobe influence, which is analogous to the one during the range compression. The following is the detailed procedures for azimuth compression [99,102,104].

6.9.1 Azimuth Matched Filtering

Let us assume that the angular carrier frequency ω_c from the range-compressed signal can be abolished by demodulation. Pick out the value at $t = t_n$ that offers maximum $|g_n(t)|$ by precise sampling and represents t as slow time s provides:

$$\hat{g}(s|x_c, R_c) = \exp[-j4\pi R(s)/\lambda].$$ (6.62)

The range function $R(s)$ can be extended as a Taylor series around $s_c = x_c/V_s$, the slow time at which the center of the radar beam signs the target, where V_s is the speed of the radar platform relative to the target point.

$$R(s) = R_c + \dot{R}_c(s - s_c) + \ddot{R}_c(s - s_c)^2/2 + \cdots \cdots. \tag{6.63}$$

The Doppler frequency can be defined as the time rate of the phase $\phi(s)$, which is formulated as:

$$\phi(s) = j4\pi R(s)/\lambda, \tag{6.64}$$

$$f_D(s) = \dot{\phi}/2\pi = -2\dot{R}(s)/\lambda, \tag{6.65}$$

$$\dot{f}_D(s) = \ddot{\phi}/2\pi = -2\ddot{R}(s)/\lambda. \tag{6.66}$$

The Doppler centroid and Doppler rate can be determined at $s = s_c$ as:

$$f_{Dc} = -2\dot{R}_c/\lambda, \tag{6.67}$$

$$f_R = -2\ddot{R}_c/\lambda, \tag{6.68}$$

In this view, one can obtain:

$$R(s) = R_c - (\lambda f_{Dc}/2)(s - s_c) - (\lambda f_R/4)(s - s_c)^2. \tag{6.69}$$

In this circumstance, the range-compressed signal can be articulated in expressions of f_{Dc} and f_R as:

$$\hat{g}(s|s_c, R_c) = \exp(-j4\pi R_c/\lambda)\exp\{j2\pi[f_{Dc}(s - s_c) + f_R(s - s_c)^2/2]\},$$

$$|s - s_c| < S/2, \tag{6.70}$$

where S is the azimuth integration time. This reveals a linear FM wave with center frequency f_{Dc} and frequency rate f_R. In this sense, the azimuth compression is to calculate the correlation as:

$$\varsigma(s|s_c, R_c) = \int_{s_c - S/2}^{s_c + S/2} h^{-1}(s' - s|s_c, R_c)\hat{g}(s'|s_c, R_c)ds'. \tag{6.71}$$

Related to range compression, the azimuth compression can be approximately grasped using a correlator function as:

$$h^{-1}(s|s_c, R_c) = \exp[-j2\pi(f_{Dc}s + f_R s^2/2)]. \tag{6.72}$$

Equation 6.72 can lead to the sequence of azimuth compression as:

$$\varsigma(s|s_c, R_c) = \exp(j2\pi f_{Dc}s)\exp(-j4\pi R_c/\lambda) \cdot \frac{\sin \pi f_R S(s - s_c)}{\pi f_R(s - s_c)} \tag{6.73}$$

Derivation of Equation 6.73 is analogous to that of range compression. The peak of this pulse occurs at $s = s_c$, the object azimuth locality [102,106,108]. Therefore, the 3 dB width of this pulse can determine the **azimuth time resolution** as:

$$\delta s = 1/|f_R|S = 1/B_D \tag{6.74}$$

where $B_D = |f_R|S$ is the Doppler bandwidth. The azimuth spatial resolution (or, azimuth **resolution**) is then computed as:

$$\delta x = V_s \delta s = V_s / B_D = V_s / |f_R| S.$$ (6.74)

From the simple geometry of a radar antenna, with a physical length L_a along-track, the nominal beam width is $\theta_H = \lambda / L_a$ so that any particular object point at the range R_c is illuminated for a nominal time $S = \lambda R_c / V_s L_a$. Based on a squint angle θ_s (Fig. 6.22), the Doppler parameters can be geometrically clarified as:

$$R^2(s) = R_c^2 + V_s^2(s - s_c)^2 - 2R_c V_s(s - s_c)\sin\theta_s$$ (6.75)

$$R(s) \approx R_c + V_s^2(s - s_c)^2 / 2R_c - V_s(s - s_c)\sin\theta_s$$ (6.76)

$$f_{Dc} = (2V_s / \lambda) \sin\theta_s,$$ (6.77)

$$f_R = -2V_s^2 / \lambda R_c.$$ (6.78)

In the circumstance of a real aperture radar, azimuth resolution swaps inversely with the physical length of its antenna as $\delta x = \lambda R_c / L_a$. For SAR, the Doppler bandwidth is, therefore, $B_D = 2V_s / L_a$ and the system azimuth resolution is determined as:

$$\delta x = L_a / 2 \cdot$$ (6.79)

where δx changes in proportion to the physical length of SAR antenna [103,106,108]. With intensive signal compression processing, high azimuth resolution can be accomplished with a SAR in a considerably minor physical dimension than that in a real aperture radar as discussed earlier.

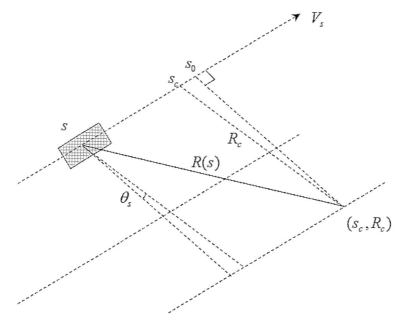

Figure 6.22. Doppler geometry parameters.

6.10 Speckles

Speckle is basically a form of noise, which degrades the quality of an image and may make its visual or digital interpretation more complex. A speckle pattern, consequently, is a random intensity pattern fashioned by the mutual interference of a set of wavefronts having different phases. Under this circumstance, they add together to give a resultant wave whose amplitude, and therefore intensity, varies randomly. In this context, if each wave is modeled by a vector, then it can be seen that a number of vectors with random angles are added together. The length of the resulting vector, therefore, can be anything from zero to the sum of the individual vector lengths—a 2-dimensional random walk (Fig. 6.23), sometimes known as a drunkard's walk. Further, when a surface is illuminated by a microwave spectra, according to diffraction theory, each point on an illuminated surface acts as a source of secondary spherical waves. The microwave spectra at any point in the scattered microwave field are made up of waves which have been scattered from each point on the illuminated surface. If the surface is rough enough to create path-length differences exceeding one wavelength, giving rise to phase changes greater than 2π, the amplitude, and hence the intensity, of the resultant backscatter microwave vary randomly [109–119].

In general, all radar images appear, to some degree, what we call radar speckle. **Speckle** appears as a grainy "salt and pepper" texture in an image (Fig. 6.24). This is produced by random constructive and destructive interference from the multiple scattering returns (Fig. 6.25) that will occur within each resolution cell [112]. Constructive interference is an increase from the mean intensity and produces bright pixels. In contrast, destructive interference is a decrease from the mean intensity and produces dark pixels (Fig. 6.26).

As an example, a homogeneous target, such as a large grass-covered field, without the effects of speckle would generally result in light-toned pixel values in an image. However, reflections from the individual blades of grass within each resolution cell result in some image pixels being brighter and some being darker than the average tone, such that the field appears speckled. The high speckle noise in SAR images, however,

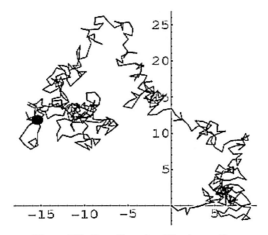

Figure 6.23. Two-dimensional Random walks.

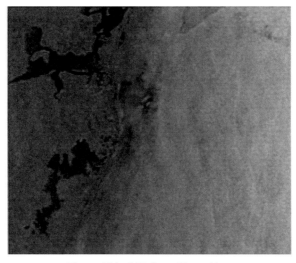

Figure 6.24. SAR data with speckles.

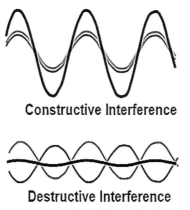

Constructive Interference

Destructive Interference

Figure 6.25. Constructive and destructive interferences.

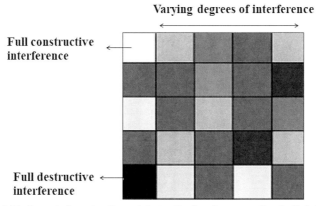

Figure 6.26. Impact of constructive and destructive interference on the pixel's brightness.

has posed great difficulties in inverting SAR images for mapping morphological features. Speckle is a result of coherent interference effects among scatterers that are randomly distributed within each resolution cell [113]. The speckle size is a function of the spatial resolution, which induces errors in the morphological feature signature detections. In order to reduce these speckle effects, appropriate filters, e.g., Lee, Gaussian, etc. [112], can be used in the preprocessing stage. The effectiveness of these speckle-reducing filters, nevertheless, is influenced by local factors and application. In fact, all speckles in SAR images are related to local changes on the Earth's surface roughness.

6.11 SAR Satellite Sensors

Satellites with Synthetic Aperture Radar (SAR) orbit the Earth in a sun-synchronous, i.e., "**L**ow **E**arth **O**rbit"; LOE any orbit around the Earth with an orbital period of less than about two hours. LEO polar orbit and data acquisitions can be made at any time of day or night and independent of cloud coverage, collecting both amplitude and phase data. The SAR satellites have the repeating paths which, using two-phase datasets for the same location at different times, allows for Interferometric SAR (InSAR) showing relative ground displacements between the two datasets along the direction of the radar beam. The SAR satellites operate at a designated frequency with L-band, C-band, and X-band being the predominate wavelengths. Below is a chart of past, present and projected SAR satellite missions.

Various agencies support the different SAR missions (Table 6.1):

- European Space Agency (ESA): ERS-1, ERS-2, Envisat, Sentinel-1.
- Japan Aerospace Exploration Agency (JAXA): JERS-1, ALOS-1, ALOS-2.
- Canadian Space Agency (CSA): Radarsat-1, Radarsat-2, Radarsat constellation.
- Deutsches Zentrum für Luft- und Raumfahrt e.V. (DLR): TerraSAR-X, TanDEM-X.
- Indian Space Research Organization (ISRO): RISAT-1, NISAR (w/ NASA).
- Comision Nacional de Actividades Espaciales: SAOCOM.
- Italian Space Agency (ASI): COSMO-Skymed.
- Instituto National de Técnica Aeroespacial (INTA): PAZ.
- Korea Areospace Research Institute (KARI): KOMPSat-5.
- National Aeronautics and Space Administration (NASA): NISAR (w/ISRO) [115].

Since the launch of Seasat in 1978, there has been a fixed development of civilian SAR satellites in space. These contain SIR-A, SIR-B, SIR-C/X-SAR, ERS-1/2, JERS-1, ENVISAT ASAR, RADARSAT-1&2, SRTM, COSMO-SkyMed, TerraSAR-X, and Advanced Land Observing Satellite (ALOS)/PALSAR. Nowadays, there are a quantity of SAR constellations and satellite platforms offering a variety of SAR data with more scheduled for the near future ensuring a permanent resource of spaceborne SAR data for oceanography applications (Table 6.1). Conversely, most of these SAR sensors deliver a variety of beam modes from 1 m spotlight modes (with limited swath coverage) to low spatial resolution ScanSAR modes (50–100 m

Table 6.1. List of SAR Satellite Sensor Characteristics.

Satellites	Acquisition period	Band Frequency	Polarization mode	Spatial resolution (m)	Revisit time (days)	Scenecover (km)
ERS-1/2	91–11	C	VV	20	35	185 × 185
JERS	92–98	L	HH	20	44	75 × 75
RADARSAT	95–13	C	HH	10–100	24	35 × 500
ASAR	1–13	C	1 or 2 pol. HH/HV/VV	30–1000	Few-35	100 × 500
PALSAR	7–11	L	Polarimetric HH/HV/VV	10–100	Few-24	100–500
RADARSAT-2	2007–	C	Polarimetric HH/HV/VV	1–15	5–10	NA
TerraSAR-X	2007–	X	1 or 2 pol. HH/HV/VV	1–20	Few-11	5–100
Cosmo-Skymed	2007-	X	1 or 2 pol HH/HV/VV	1-100	12	10–200
SAOCOM	2015–	L	Polarimetric HH/HV/VV	7–100	Few-16	60–320
Sentinel1	2015	C	1 or 2 pol HH/HV/VV	5–100 m	Few-12	80–400
ALOS-2	2015	L	Polarimetric HH/HV/VV	3–100	Few-14	25–350

resolution) with large swath coverages. In this regard, each of these SAR satellite sensors has a web page or other information sources, where detailed information on the sensor specifics, data processing and products can be acquired [24].

6.12 Waves and Frequency Ranges Used by Radar

Radar systems work in an extensive band of transmitting frequencies. The greater the frequency of a radar system, the greater it is affected by climate stipulations such as rain or clouds. But the higher the transmitted frequency, the higher is the accuracy of the radar system. Here the different radar frequency bands are discussed.

6.12.1 A and B(HF-und VHF) Radar Bands

A and B(HF-und VHF) radar bands are below 300 MHz, which have a long conventional use because these frequencies signified the frontline of radio technology at the time during the World War II. Nowadays, these frequencies are exploited for primary warning radars and termed "Over The Horizon (OTH) Radars" (Fig. 6.27). By means of these lower frequencies, it is straightforward to acquire high-power transmitters. The reduction of the electro-magnetic waves is lower than using higher frequencies. On the other hand, the precision is restricted, since a lower frequency necessitates antennas with a precise large physical size, which regulates angle precision and angle resolution. In this view, these frequency-bands are exploited by other communications

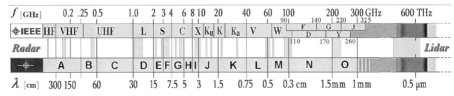

Figure 6.27. Wavelength and frequency ranges of radar.

and broadcasting services too. Consequently, the bandwidth of the radar is imperfect specially at the expense of precision and resolution again [108,117].

6.12.2 C-Band (UHF-Radar)

There are particular radar sets established for this frequency band (300 MHz to 1 GHz). C-Band is a reputable frequency for the operation of SAR for the detection and tracking of satellites and ballistic missiles over a long range. These radars function for initial warning and target acquisition corresponding the surveillance radar for the Medium Extended Air Defense System (MEADS). Nearly all weather radar-applications, especially wind profilers function with these frequencies since the electromagnetic waves are identical low and exaggerated by clouds and rain. Therefore, the new technology of Ultrawideband (UWB) radars operates all frequencies from A- to C-Band. UWB-radars transmit very low pulses in all frequencies concurrently. They are utilized in the precise material investigation and for example, Ground Penetrating Radar (GPR) for archaeological explorations [117].

6.12.3 D-Band (L-Band Radar)

This frequency band (1 to 2 GHz) is favored for the operation of long-range air-surveillance radars out to 250 NM ($\approx$ 400 km). They convey pulses with high energy, broad bandwidth and an intrapulse modulation regularly. Owing to the curvature of the earth the realizable maximum range is restricted for targets hovering with low altitude. In fact, these targets vanish incredibly fast behind the radar horizon. Conversely, in Air Traffic Management (ATM) long-range surveillance radars similar to the Air Route Surveillance Radar (ARSR) operates in L-frequency band. Coupled with a Monopulse Secondary Surveillance Radar (MSSR) they use a relatively large, but the slower rotating antenna. The designator L-Band is good as mnemonic rhyme as large antenna or long range [109].

6.12.4 E/F-Band (S-Band Radar)

The atmospheric attenuation is more complex than in D-Band. Radar sets are a necessity for significantly higher spreading power than in lower frequency ranges to accomplish a virtuous maximum range. For instance, the Medium Power Radar (MPR) operates with a pulse power of up to 20 MW. In this frequency range the impact of weather circumstances is greater than in D-band. Thus a combine of weather radars work on

E/F-Band, however, further in subtropic and tropic climatic environments, since the radar can see beyond a harsh storm.

In this sense, Special Airport Surveillance Radars (ASR) are depleted at airports to identify and display the location of an aircraft in the terminal zone with a medium range up to 50…60 NM (≈ 100 km). An ASR senses the aircraft location and weather situations in the area of civilian and military airfields. The designator S-Band (contrary to L-Band) is of high quality as mnemonic rhyme as smaller antenna or shorter range [117].

6.12.5 G-Band (C-Band Radar)

In G- Band there are numerous mobile military battlefield surveillance, missile-control and ground surveillance radar sets with a short or medium range. The dimension of the antennas delivers an exceptional precision and determination, nonetheless, the relatively small-sized antennas do not perturb a fast moving one. The impact of imperfect weather circumstances is incredibly extraordinary. Thus, air-surveillance radars deplete an antenna often maintained with circular polarization. This frequency band is encoded for furthermost sorts of weather radar, which operate to locate precipitation in temperate zone such as Europe.

6.12.6 I/J-Band (X- and Ku-Band Radars)

In this frequency-band (8 to 12 GHz) the connection between castoff wavelength and dimension of the antenna is significantly improved than in lower frequency-bands. The I/J-Band is a comparatively popular radar band for military applications such as airborne radars for execution of the roles of interceptor, fighter and attack of enemy fighters and of ground targets. A precise small antenna size delivers a high quality operation. Missile guidance systems at I/J-band are of a convenient dimension and are, hence, of interest for submissions where suppleness and tiny weight are significant and precise long range is not a main prerequisite.

This frequency band has been depleted widespreadly for maritime civil and military navigation radars. Identical small and cheap antennas with a high rotation speed are tolerable for a fair maximum range and a decent precision. Slotted waveguide and small patch antennas are used as radar antenna, under a protective radome mostly. In this regard, this frequency band is also popular for spaceborne or airborne imaging radars based on Synthetic Aperture Radar (SAR) both for military electronic intelligence and civil geographic mapping. A special Inverse Synthetic Aperture Radar (ISAR) is in use as a maritime airborne instrument of pollution control [108–118].

6.12.7 K-Band (K- and Ka-Band Radars)

The higher the frequency, the higher is the atmospheric absorption and attenuation of the waves. Otherwise the achievable accuracy and the range resolution rise too. Radar applications in this frequency band provide short range, very high resolution and high data renewing rate. In ATM these radar sets are called Surface Movement

Radar (SMR) or Airport Surface Detection Equipment (ASDE). Using of very short transmitting pulses of a few nanoseconds enables a range resolution, of the aircraft that can be seen on the radar's display [108,118].

6.12.8 V-Band

By the molecular dispersion (here this is the influence of the air humidity), this frequency band stays for a high attenuation. Radar applications are limited for a short range of a couple of meters here [119–121].

6.12.9 W-Band

W-band has dual visible phenomena (i) a maximum of attenuation at around 75 GHz and (ii) a relative minimum at about 96 GHz. Both frequency ranges are in use practically. In automotive engineering small built in radar sets operate at 75…76 GHz for parking assistants, blind spot and brake assists. The high attenuation (here the influence of the oxygen molecules O_2) enhances the immunity to interference of these radar sets. There are radar sets operating at 96 to 98 GHz as laboratory equipments yet. These applications give a preview for the use of radar at extremely high frequencies of 100 GHz [121].

The exclusive designations for Radar-Frequency Bands are very confusing. This is no trouble for a radar engineer or technician. These expert scientists can deal with these exceptional bands, frequencies and wavelengths. But they are not accountable for procurement logistics, e.g., for buying of maintenance and measurement devices or even to purchase a new one radar. Unfortunately, the administration of logistics has mainly graduated in commercial enterprise sciences. Therefore, they will have trouble with the complicated band designators. The problem is now to assert, that a frequency generator for I and J-Band serves an X- and Ku-Band Radar and the D-Band Jammer interferes an L-Band Radar [119,121].

Novel Relativity Theories of Synthetic Aperture Radar

In continuation with the earlier chapters, it is well known that the radar is defined as RAdio Detection And Ranging. Ranging is the distance from the radar where the microwave photons travel across space and time and interact with the objects. In this regard, space-time is the keystone of the relativity theory. The main question that can be raised is whether SAR can be explained by relativity theories? This chapter is devoted to finding the correlation between SAR and relativity theory.

7.1 What is Simple Definition of Relativity?

Einstein proclaimed that all objects in the universe are always traveling through spacetime at one fixed speed—that of light. This is a strange idea; we are used to the notion that objects travel at speeds considerably less than that of light. We have repeatedly emphasized this as the reason relativistic effects are so unfamiliar in the everyday world. All of this is true. We are presently talking about an object's combined speed through all four dimensions—three space and one time—and it is the object's speed in this generalized sense that is equal to that of light. To understand this more fully and to reveal its importance, we note that like the impractical single-speed car discussed above, this one fixed speed can be shared between the different dimensions—different space and time dimensions. If an object is sitting still (relative to us) and consequently does not move through space at all, then in analogy to the first runs of the car, all of the object's motion is used to travel through one dimension—in this case, the time dimension. Moreover, all objects that are at rest relative to us and to each other move through time—they age—at exactly the same rate or speed. If an object does move through space, however, this means that some of the previous motion through time must be diverted. Like the car traveling at an angle, this sharing of motion implies that the object will travel more slowly through time than its stationary counterparts since some of its motion is now being used to move through space. That is, its clock will tick more slowly if it moves through space. This is exactly what we found earlier. We now see that time slows down when an object moves relative to us because this diverts some of its motion through time into motion through space. The speed of an object through space is thus merely a reflection of how much of its motion through time is diverted.

Note that Greene says—If an object is sitting still (relative to us), and it's the relative to us that is the key phrase. In this context stationary means stationary with respect to the observer making the measurements. There is no absolute meaning for stationary. Indeed, Greene spends much of the preceding part of Chapter 2 explaining that all motion is relative.

To really understand relativity (both flavors) you need to appreciate the importance of coordinate systems. If I want to measure positions, times, velocities, etc. I need to establish a measurement system. I need to choose an origin, which is conventionally at my position, and I need to choose x, y, z and time axes. This constitutes my coordinate system, and having done this I can use my rulers and clocks to measure the positions of events in spacetime and thereby measure velocities.

When we talk about time dilation or length contraction we don't mean anything funny is happening to my lengths x, y and z or my times t. What we mean is that if I measure my elapsed times t2−t1 they won't necessarily be the same as the measurements made by a different observer using a different coordinate system.

When we talk about a stationary object we mean one that shares my coordinate system, i.e., I and the object will agree on the coordinate system we use to make measurements. Obviously, this means our measurements of time and space will be the same since we're using the same coordinates.

In special relativity, a coordinate system fills all of spacetime, i.e., my axes carry on it a straight line forever. In general relativity, life is much more complicated because the curvature of spacetime means my coordinates are only good locally.

7.2 Relativistic of SAR Doppler

The principle of relativity commonly involves dual interrelated theories through Albert Einstein: special relativity and general relativity. Special relativity applies to fundamental particles and their interactions, describing all their physical phenomena barring gravity. General relativity, however, describes the regulation of gravitation and its relation to different forces of nature. It implements to the cosmological and astrophysical realm, which includes astronomy.

The specular reflection of microwave photons from a moving oil spill covered sea surface has been studied in many different contexts during the last decades. In contrast to the conventional obstruction of specular reflection from a stationary target, the situation with the moving target provides an enlargement to numerous significant differences from the conventional circumstance. Indeed, a precise curious singularity in quantum field theory that contains a moving object, which has been investigated experimentally, which is identified as a dynamical Casimir effect, in which an object experiencing relativistic motion transforms virtual photons into exactly recognizable real photons [122].

Here we will deliberate two possessions concerning specular reflection from a regularly moving sea state due to the oil covered. In this view, the angle of incidence is no longer equivalent to the angle of backscattering [123]. Further, the Doppler effect upon backscattering from the dynamical sea-state, or a change in the frequency of the microwave photon backscattering with respect to the incident photon signal [124]. Using the conservation laws of governing elastic collisions to investigate the collision of the photon with the dynamical sea-state.

Let us consider $\hbar\omega$ presents a single photon of energy and $\hbar\omega c^{-1}$ presents a corresponding momentum of magnitude, upon an oblique incident beam on dynamical sea surface oil covered moving at a speed C (Fig. 7.1) Consistent with the quantum depiction of the procedure of backscattering, the microwave photon will be reflected by the target atoms from the oil-covered sea surface. The dynamical motion of the sea-state will instigate loss or gain of momentum (and, consequently, energy) of the photon upon backscattering, relying on whether the sea surface is propagated along the positive or the negative direction of the azimuth direction (x). Let us consider that the motion of the sea-state is non-relativistic ($C \ll c$), and that the backscattering of the photon at its surface is perfectly elastic. In this understanding, the photon is treated as a spherical particle which collides and bounces-off a frictionless plane surface [125].

As a result of the interaction between the photon and the dynamical sea surface, there will be a shift in the frequency of the photon after the backscattering, and the incident angle α and the reflected angle β will not be equal. Let $\hbar\omega'$ and $\hbar\omega'c^{-1}$ signify the particular energy and the momentum magnitude of the backscattered photon. Let us implement the circumstance for the conservation of momentum along azimuth (x) and range (y) direction, then we obtain:

$$\frac{\hbar\omega}{c}\cos\alpha + MC = -\frac{\hbar\omega'}{c}\cos\beta + M(C + \Delta C_x), \tag{7.1}$$

$$\frac{\hbar\omega}{c}\sin\alpha = \frac{\hbar\omega'}{c}\sin\beta, \tag{7.2}$$

where M is the mass of the seawater which is ∞ [121–123]. Equation 7.1 demonstrates the velocity of the sea surface is reformed after the backscatter by an amount of ΔC_x in

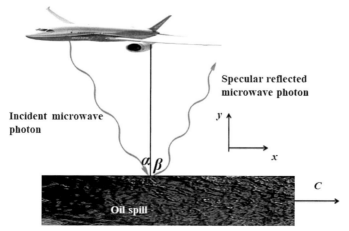

Figure 7.1. Specular reflection of a photon from oil covered.

the azimuth direction. The scientific explanation of the law of conservation of energy can mathematically be written as:

$$M \Delta C_x = \hbar c^{-1} \left(\omega \cos \alpha + \omega' \cos \beta \right),$$
(7.3)

Then the Doppler frequency shift ω' is determined from:

$$\omega' = \omega \left(1 - C \cos \alpha c^{-1} \right) \left(1 + C \cos \beta c^{-1} \right)^{-1}$$
(7.4)

Based on Equation 7.4, Equation 7.2 can be modified as:

$$\sin \beta \left(1 - \frac{C}{c} \cos \alpha \right) = \sin \alpha \left(1 + \frac{C}{c} \cos \beta \right),$$
(7.5)

Equation 7.5 can be revealed for the backscattering angle β in terms of the angle of incidence α and the velocity C of the moving sea surface to deliver:

$$\cos \beta = -2Cc^{-1} + \left(1 + C^2 c^{-1} \right) \cos \alpha \left[1 - 2Cc^{-1} \cos \alpha + C^2 c^{-1} \right]^{-1}$$
(7.6)

The changing in the photon behavior before and after backscattering can be obtained by substitution of Equation 7.6 into 7.4:

$$\omega' = \omega \left(1 - 2C \cos \alpha c^{-1} + C^2 c^{-2} \right) \left(1 - C^2 c^{-2} \right)^{-1}$$
(7.7)

The backscatter photon in Equations 7.6 and 7.7 matches with the formulas achieved by means of relativistic electrodynamics, the principles of wave optics or directly from the postulates of special relativity [3,7,12,14–17,19–21]. The remarkable fact is that in developing the similar equations, we merely expend the associations for the energy and momentum of a microwave photon in the sense of the photon model, which were deceptively revealed and established independently of special relativity [22]. Another curious argument is that the resulting Formulas 7.6 and 7.7 are essentially nonrelativistic theory, which is also effective in the circumstance when the sea surface is moving at relativistic speeds [122–126].

Generally, the relativistic Doppler impact is the exchange in frequency (and wavelength) of light, triggered through the relative motion of the SAR antenna and the observer (as in the classical Doppler effect), when taking into account consequences described through the distinctive principle of relativity [124]. The relativistic Doppler impact is unique from the non-relativistic Doppler impact as the equations encompass the time dilation impact of special relativity and do no longer contain the medium of propagation as a reference point. They describe the complete distinction in determining frequencies and possess the required Lorentz symmetry [126].

7.3 Time Dilation

From the point of view of the relativity theory, time dilation is a distinction in the elapsed time measured through dual observers, either due to a speed distinction relative to each other or through being otherwise located relative to a gravitational field. In the consequence of the nature of space-time, a time that is shifting relative to a receiver will be measured to impulse slower than a time that is on relaxation in the antenna's frame of reference. In this regard, a period that is underneath the impact of a superior gravitational field than an antenna will additionally be measured to tick slower than the time of radiated beams from the antenna. This is known as a delay time and symbolizes as τ (sec). The scientific explanation of the time interval between two pulses $\Delta\tau$ in that SAR frame can mathematically be given by:

$$\Delta\tau = -\Delta t \sqrt{1 - \frac{v^2}{c^2}} \qquad (7.8)$$

where v present SAR beam velocity, c is the speed of light and Δt is time dilation effect. For instance, for GPS satellites to operate, they should modify for comparable bending of space-time to coordinate with systems on Earth [127]. In other words, timers on the Space Shuttle and SAR is walking a little slower than reference clocks on Earth, or clocks on GPS and Galileo satellites running barely faster.

7.4 Length Contraction

Suppose now that the length of oil covered L_0 is imagined in the SAR's inertial frame (Fig. 7.2). Let us presume that there is a far away SAR beam source in the same line that oil covered is poignant, emitting microwave pulses with frequency f_0, and that oil covered's acceleration and deceleration times are short, equated to the entire traveling time. Primarily we should memorandum that each microwave pulse that passes through the oil-covered will necessarily reach the SAR antenna before sending the other sequence pulses, since $v < c$. In spite of the sea surface fluctuations, the specular reflection of the oil-covered still can be received by the SAR antenna. Let us consider the entire travel time measured by SAR is $2Rv^{-1}$, and so the number of pulses received by the SAR antenna is given by:

$$N = 2Rv^{-1} \qquad (7.9)$$

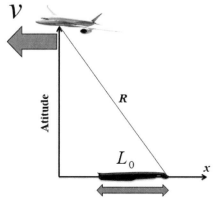

Figure 7.2. Oil covered length contraction effect in the SAR image.

The Doppler formula must be used to determine, for instance, the number of pulses that hit the oil covered. With this in view, imagine oil covered at the point x_0 on SAR, the SAR antenna receiver will:

$$N_+ = f_0 \left(\frac{\left(1 + vc^{-1}\right)}{\sqrt{1 - vc^{-1}}} \right) L' v \tag{7.10}$$

where f_0 is SAR frequency, v is SAR velocity in orbit and L' is an unknown distance in which SAR is imagined at the point x_0. In this regard, the length contraction effect (L') is then formulated as:

$$L' = L_0 \sqrt{1 - v^2 c^{-2}} \tag{7.11}$$

The length of oil covered in SAR images will experience the length contraction effect as demonstrated in Equation 7.11. In other words, the length of any object in a moving SAR frame will appear foreshortened in the direction of motion or contracted. This contraction (also known as Lorentz contraction or Lorentz–Fitz Gerald contraction after Hendrik Lorentz and George Francis Fitz Gerald) is usually only noticeable at a substantial fraction of the speed of light. Length contraction is only in the direction in which the body is traveling. For standard objects, this effect is negligible at everyday speeds and can be ignored for all regular purposes, only becoming significant as the object approaches the speed of light relative to the observer. In general, this contraction only occurs along the line of SAR motion, i.e., azimuth direction, and Equation 7.11 is simplified as:

$$L' = \frac{L_0}{\gamma(v)} \tag{7.12}$$

where $\gamma(v)$ is the Lorentz factors, which is the factor by which time, length and relativistic mass change for an object while that object is moving. The inverse term $\frac{1}{\sqrt{1 - v^2 c^{-2}}}$ in Equation 7.11 presents Lorentz factors. It is clear that the inverse of Lorentz factors is a function of velocity and leads to a circular arc (Fig. 7.3).

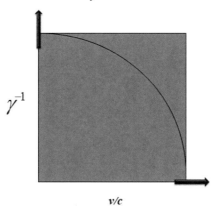

Figure 7.3. The circular arc of inverse Lorentz factors.

Furthermore, an inertial SAR reference frame in relative motion with the moving frame of the oil-covered dynamic sea-state measures a period dilated by the γ Lorentz factor. If one wants to compute the time dilation in the oscillator period of oil-covered as imaged by the SAR moving frame, we must consider γ factor, which is attached to the rest mass of the oil-covered and a second γ factor inversely attached to the dynamic of the sea-state. The latter is due to the definition of the four-force in special relativity. Needless to say, moving objects appear to experience the length contraction because they are perceived in space-time cross-section. In this sense, the length contraction is considered as some sort of optical illusion depending on the sensor. It is only meant to emphasize that length contraction appears differently in various SAR frames, as verified by reasonable coordinates.

In this understanding, the object geometry is only the one interpreted, especially if the object is deformed: because the changing object will actually appear rotated-and-stretched-larger by the Lorentz boost in the space-time diagram, and it is only after a non-perpendicular cross section is considered that this from γ down to $1/\gamma$ is corrected. In fact, object x-coordinate is elongated by a factor of γ; the Euclidean distance should be $\gamma\sqrt{1+v^2c^{-2}}$.

7.5 Does SAR Polarization Cause Length Contraction?

The SAR sensor transmits a longitudinal electromagnetic wave. It is possible to transmit the longitudinal wave in a single plane (polarization). Usually two polarization are used: (i) Horizontal (H) and (ii) Vertical (V) (Fig. 7.4).

Conversely, it is conceivable for the SAR sensor to pick out the polarization of the received signal. Usually, most scatterers reflect the wave in the same polarization (co-polarized: HH, VV). Nonetheless, some of the signals may backscatter in a dissimilar plane (cross-polarized: HV, VH). In this case, HH presents a horizontal transmit and horizontal receive (Fig. 7.5), while HV stands for horizontal transmit and vertical receive (Fig. 7.6). However, VH involves vertical transmit and horizontal receive (Fig. 7.7), while VV presents vertical transmit, and vertical receive (Fig. 7.8).

Both the co-polarized signal and cross-polarized signal have different characteristics. The co-polarized signal is usually strong and causes specular, surface and volume scattering. However, the cross-polarized signal is usually weak, associated with multiple scattering and has strong relationships with the orientation of scatters. Thus, the polarized signal regularly relies on the incidence angle due to the strong

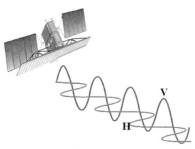

Figure 7.4. SAR two polarization H and V.

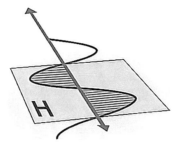

Figure 7.5. HH polarization.

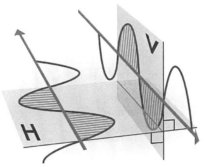

Figure 7.6. HV polarization

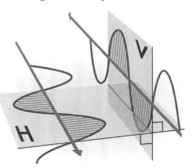

Figure 7.7. VH polarization.

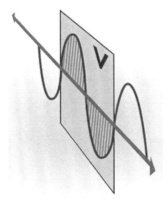

Figure 7.8. VV polarization.

association with geometry. In this regard, VV polarization is used extensively for studying capillary waves on the surface of water bodies while cross-polarization is useful when the volume (multiple) scattering occurs. It can allow the separation of soil and vegetation and of forest and non-forest.

Consistent with the above perspective, the VV polarization provides better results than the other polarizations for detection and monitoring oil spill scenarios than VH, HV, and HH polarizations. From the point of view of length contraction, the size and length of the oil spill have to be imagined differences among a variety of polarization. Figure 7.9 demonstrates the length contraction for oil spill in VV polarization (Fig. 7.9) and quadrature-polarimetric (quad-pol) Synthetic Aperture Radar (SAR) systems (Fig. 7.9b), which allow the measurement of the 2×2 complex scattering matrix S. The scattering matrix S is usually built up by operating the radar with the interleaved transmission of alternate H- and V-polarized pulses and simultaneous reception of both H and V polarizations (conventional quad-pol SAR). It is clear that length contraction takes place in the quad-pol SAR data (Fig. 7.10) as the oil spill geometry and wavelength are totally constricted.

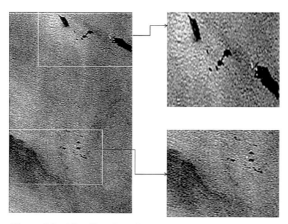

Figure 7.9. Oil spill and swell wavelength in VV polarization.

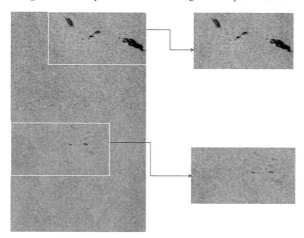

Figure 7.10. Oil spill and swell wavelength in quadrature-polarimetric.

The length contraction occurs clearly in quad-pol SAR image due to its narrow swath width. This restricts the application of routine mapping of marine oil spills, causing more difficulty in large-scale sea surface monitoring.

7.6 SAR Time and Range Relativities

Let us assume that SAR traveling on orbit $O: x^\mu = x_O^\mu(t)$ with the appropriate time t. In this case, we can establish the SAR parameters, which can lead to the special relativity as:

$t^+(x) \equiv$, which presents proper time at which a microwave photon beam, i.e., technically, a null geodesic, leaving point x could intercept O. However, $t^-(x) \equiv$ presents proper time at which a microwave photon beam (null geodesic) could leave O, and still reach a point x. In this regard, the SAR time is formulated as:

$$t\ (x) \equiv 0.5\big(t^+(x) + t^-(x)\big) \tag{7.13}$$

Therefore, the SAR distance R is expressed as;

$$R(x) \equiv 0.5\big(t^+(x) + t^-(x)\big) \tag{7.14}$$

The new definition of SAR 'hypersurface of simultaneity $\sum_{t_0}$ at a time t_0 is introduced as (Fig. 7.11):

$$\sum_{t_0} \equiv \{x : t(x) = t_0\} \tag{7.15}$$

This formula simply articulates that SAR can assign a time to a distant event by sending a microwave signal to the objects and backscatter and averaging the (proper) times of sending and receiving. Nonetheless, the concept of delineating hypersurfaces of simultaneity in terms of 'SAR time' has rarely been applied to noninertial satellite orbit or to non-flat spacetimes. This is perhaps due to Bondi's claim [133] that "how a clock reacts to acceleration is utterly dependent on how the clock is constructed". Consistent with Dolby [134] the postulation that 'proper times' will show identically

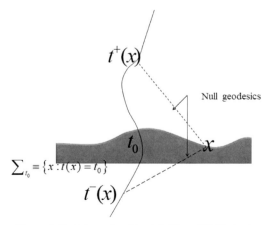

Figure 7.11. Schematic of the definition of 'SAR time'.

as a function of acceleration or as a function of gravitational fields, which is a simple principle of general relativity, without which proper time would have no physical meaning. There is, therefore, no purpose to use 'radar time' to speed up SAR or to warp space-times. Since SAR time can be described short of any remark of coordinates it is, by construction, independent of our choice of coordinates. In this case, radar time or SAR time is single-valued, corresponds to proper time on the SAR path, and is invariant as a function of 'time-reversal'—that is, under reversal of the sign of the SAR's proper time [134].

7.7 The Relativity of Frequency Changing

There are essentially dual general relativistic impacts that can produce systematic sources of inaccuracy in the Newtonian reduction of SAR observations. The initial is the general relativistic Doppler shift which was discussed earlier. Conversely, the difference between the gravitational potentials of the transmitting and receiving radars (in the lowest order post-Newtonian approximation) is proportional to the amplitude of the change. In this view, the shift in wavelength $\Delta\lambda$ which is corresponding to the frequency shift Δf is formulated as:

$$\frac{\Delta\lambda}{\lambda} = \Delta\phi c^{-2} \tag{7.16}$$

here ϕ is the negative of the usual Newtonian gravitational potential which equals $\phi \simeq GM_E R_E^{-1}$. However, this manifestation discounts both the oblateness of the Earth and third-body perturbations. Let us consider third-body perturbations owing to an object of mass M at a geocentric distance D. The mathematical expression of change in the gravitational potential due to the effects of the Earth's oblateness can be much larger and is given by

$$c^2 \frac{\Delta\lambda}{\lambda} \leq 2GM_E \frac{J_2}{R_E^3} \tag{7.17}$$

$\frac{\Delta\lambda}{\lambda} \leq 3.3 \times 10^{-9}$ which is meaningfully superior to the second order special relativistic effect owing to the Earth's rotation, i.e., $\omega_E = 15''.041 \sec^{-1}$, $R_E = 6378.1 km$, and J_2 is the geopotential which equals 0.00108.

The succeeding general relativistic impact is caused by the deficiency of flatness (in Minkowski space) of the general relativistic line element near massive bodies. These impacts in a departure of the speed of light near such an object from its speed. Hereafter space for an object, as inferred by splitting the round trip flight time and then dividing by c, must be adjusted. Therefore, in the lowest order post-Newtonian approximation, the quantity of the rectification is equivalent to the path integral of the (Newtonian) gravitational potential [135]. Thus

$$\Delta_s \simeq \left(2GM_E c^{-2}\right) \ln\left(R_{sat} R_E^{-1}\right) \tag{7.18}$$

Equation 7.18 reveals the amounts of Δ_s to be 1.7 mm for a near-stationary satellite. Conversely, there are notwithstanding dualistic approximations within the context of a special relativistic computation that requires a numerical analysis. The first involves the equivalent between the transmitted pulse P_T and received one P_R, i.e., $P_T = P_R$.

As stated above, let first assume the velocity approximation when the transmitting and receiving radars are colocated. Then the difference ΔP between the transmitted pulse P_T and received one P_R is the result of a change of direction owing to the rotation of the Earth during the pulse's travel time. In this regard, $\Delta P = \omega_E \Delta t$, where $\Delta t = 10^{-5}$ radiance. Hence the $P_T = P_R$ approximation is of the order of the Earth relative velocity β_E, i.e., $10^{-5} \beta_E$.

Continuing in this anthropomorphic vein, the transmitting radar also knows the direction of propagation of the outgoing pulse as perceived in this same inertial frame (unit vector $= \underline{n}_T$). However, the receiving direction frame is not parallel to the transmitted one because they are measured in two different inertial frames (not because of the parallactic displacement of the radars). Therefore, the size of this difference is of the order of the velocity difference between the two reference frames. This difference, when the transmitting radar is colocated with the receiving radar, has just been shown to be of the order of the Earth relative velocity of $10^{-5} \beta_E$ [136].

7.8 Invariance of Space-Time Interval

Let us assume the transmitter and receiver platforms travel with the same constant speed so that their positions can be considered stationary in a platform centered reference frame. Moreover, the transmitter and receiver clocks are assumed to be perfectly synchronized within this frame, independent of the actually employed synchronization technique (Fig. 7.12). For instance, TSX and TDX represent the transmitter and receiver satellites, which are detached by a constant baseline B. In this scenario, as long the satellites are stationary, the scene objects travel reasonably to this frame with a velocity v. In this reference frame, the time interval between the transmission and reception of a radar pulse is signified by Δt.

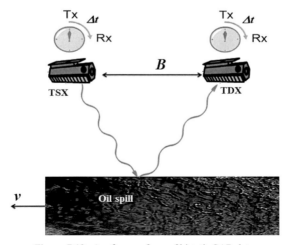

Figure 7.12. A reference form of bistatic SAR data.

With references to the special theory of Relativity, the speed of light has permanently the identical value, liberated of the inertial reference frame one practices to designate a physical system. An instantaneous significance is well known as non-simultaneity of events. In this understanding, dual spatially detached occurrences, which happen at a similar time in one reference frame, perhaps can no longer be simultaneous in another reference frame that moves relative to the first one. Consequently, radar transmitters and receivers, that are entirely synchronous in the platform centered frame, which is no longer synchronous in the Earth Centred Earth Fixed (ECEF) reference frame. The scientific explanation of the theory of relativity is the reliability of the space-time interval that can be given by:

$$\Delta s = \sqrt{\left(c.\Delta t^2\right) - \sum_{i=1}^{3} \Delta x_i^2} \tag{7.19}$$

Equation 7.19 presents the time Δt and position Δx differences, respectively, between dual occasions as observed in a given reference frame. Moreover, Δs remains invariant under the Lorentz group of linear spacetime transformations. Thus, in the platform reference frame, the space-time interval between the transmit (Tx) and receive (Rx) events is computed using:

$$\Delta s = \sqrt{\left(c.\Delta t^2\right) - \left\|\vec{B}\right\|^2} \tag{7.20}$$

Let us consider, along-track baseline between the satellites differs by $v.\left(r_{bi}c^{-1}\right)$ from that delivered in the platform centered reference frame if one compares the transmit and receive events. Here, v denotes the receiver velocity, *rbi* presents the bistatic range (Fig. 7.13).

Based on Fig. 7.13, the interval between these events is delivered by:

$$\Delta s = \sqrt{\left(c.\frac{r_{bi}}{c}\right)^2 - \left\|\vec{B} + \vec{v}.\frac{r_{bi}}{c}\right\|^2} \tag{7.21}$$

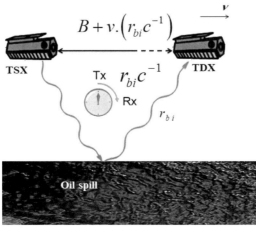

Figure 7.13. Bistatic SAR data acquisition as seen from the Earth Centred Earth Fixed (ECEF) reference frame.

Both Equations 7.20 and 7.21 are equivalent to each other as:

$$\sqrt{\left(c.\Delta t^2\right)-\left\|\vec{B}\right\|^2}=\sqrt{\left(c.\frac{r_{bi}}{c}\right)^2-\left\|\vec{B}+\vec{v}.\frac{r_{bi}}{c}\right\|^2} \tag{7.22}$$

The quantity can be determined from Equation 7.22 as:

$$r_{bi}\approx c.\Delta t+\vec{B}.\vec{v}c^{-1} \tag{7.23}$$

Equation 7.23 has indicated that the platform velocities are smaller than the speed of light. Equation 7.23 demonstrates that $c.\Delta t$ represents the product of the velocity of light with the time difference between the T_x and R_x events as described in the platform frame, where radar data acquisition and recording are achieved.

Taking into account the space-time structure of special relativity $\vec{B}.\vec{v}c^{-1}$ occurs. Conversely, $\vec{B}.\vec{v}c^{-1}$ is proportional to the scalar product between the platform velocity vector $\vec{v}$ and the baseline vector $\vec{B}$. In other words, it increases with both the along-track baseline between the satellites and the satellite velocity. For instance, a satellite tandem is flying with a velocity of 7.5 km/s and an along-track baseline of 1 km, the $\vec{B}.\vec{v}c^{-1}$ amounts to a bistatic range error of 2.5 cm. In this regard, this relativistic range offset may announce minor, nonetheless for an interferometric X-band system with a wavelength of 3.1 cm, the phase error is with 290°, which is already close to the ambiguity interval.

Consistent with Krieger and De Zan [137], the procedure of different reference frames for bistatic SAR processing and bistatic radar synchronization is predisposed to the distinguished phase and time errors. These mistakes are a direct consequence of the relativity of simultaneity and can be explained in good approximation within the framework of Einstein's special theory of relativity. Consequently, using the invariance of the space-time interval, an analytic expression is derived which shows that the time and phase errors increase with increasing along-track distance between the satellites.

Quantization of Oil Spill Imagining in Synthetic Aperture Radar

Quantum mechanics is a keystone to comprehend the mechanism of Synthetic Aperture Radar (SAR) for the imaging oil spill. Indeed, the chemical compound of an oil spill can only be explained by quantum mechanics. Most of the compounds in crude oil involve molecules made up of hydrogen and carbon atoms only, which are known as hydrocarbon compounds. In this sense, the hydrocarbon atoms have several characteristics, for instance, energy, momentum, besides wave-particle duality. For instance, hydrocarbon atom wave-particle duality is the perception of quantum mechanics that each particle or quantic entity can be partially expressed in terms not only of particles (atoms) but also of waves. In this case, if the crude oil discharges to seawater, it drifts across a fluid bath by the waves and currents. In this view, the crude oil particles will behave like waves. The critical question is, can the quantum field theory explain the behavior of oil spill in the ocean? Further, how, do quantum mechanics provide an explanation for the trajectory movement of oil spills? Finally, chemical components and behavior of oil spill covered water? This chapter is devoted to delivering modern findings and theories behind quantizing oil spills.

8.1 Quantization of Crude Oil Chemical Chains

Quantum mechanics can scrutinize a reliable quantification of crude oil as a function of its absorption spectra. Let us postulate that crude oil contains a number of carbon atoms n with edge length b and d represent the average spacing between carbons $\sim 1.4\text{A}°$. The mathematical description of crude oil as a function of its hydrocarbon chain can be written as [138]:

$$\eta(x) = 0 \text{ for } 0 \le x \le L_n$$
$$= \infty \text{ for } \text{otherwise}$$

(8.1)

Equation 8.1 represents the hydrocarbon chain in the one-dimensional 1-D infinite potential well $\eta(x)$ where the Fermi level L is defined as:

$$L = (n-1)d + 2b \qquad (8.1.1)$$

The Fermi energy is an insight in quantum mechanics regularly denoting the energy alteration between the highest and lowest (Fig. 8.1), which occupies the single-particle state in a quantum system of non-interacting fermions at absolute zero temperature. Specifically, the levels of potential energy E_k are formulated as:

$$E_k = \frac{\hbar^2 k^2}{2m(nd)^2}. \qquad (8.2)$$

here $\hbar$ is h-bar, which is a modified form of Planck's constant and equals $\frac{h}{2\pi}$. In this regard, if electrons are dominated in $n+1$, the ground state energy E_0 of the hydrocarbon is calculated using:

$$E_0 = \frac{\hbar^2 \pi^2}{24m(nd)^2}(n^3 + 3n^2 + 5n + 3). \qquad (8.3)$$

Equation 8.3 demonstrates that the energy of the first excited state can be computed by projecting an electron from the Fermi level to the next excited state as:

$$E_1 = \frac{\hbar^2 \pi^2}{24m(nd)^2}(n^3 + 3n^2 + 17n + 27). \qquad (8.4)$$

The wavelength of light absorbed in a ground state-to-first excited state transition is a function of the Fermi level, i.e., the difference between E_0 and E_1 as proportional inversely with the absorbed wavelength λ_n is determined from [139]:

$$\Delta E = \frac{hc}{\lambda_n} \qquad (8.5)$$

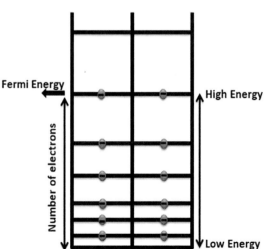

Figure 8.1. The Fermi energy concept.

In this view, the wavelength of light absorbed λ_n by crude oil or hydrocarbon atoms is computed by:

$$\lambda_n^N = \frac{\lambda_n^0}{1+(-1)^{\frac{n+3}{2}}\gamma\frac{n}{n+2}} \tag{8.6}$$

where λ_0 is the wavelength of absorbed light at the first Femi level, and γ is given by:

$$\gamma \equiv \frac{4m\alpha\eta_0 d}{\hbar^2\pi^2} \tag{8.6.1}$$

here α energy eigenvalues. Equation 8.6 explains the level of the light wave spectra absorption by oil crude. In this understanding, it was documented that crude oil molecular ions with different numbers of carbon atoms absorb a different wavelength of electromagnetic spectra. In this view, the crude oil, with $n \geq 850nm$ or $n \leq 350nm$ does not absorb colored light. In this circumstance the existing nitrogen atoms in the center of hydrocarbon spectra, which is red or blue wavelength, shift contingent on whether $\frac{n+1}{2}$, where $n + 1$ is the number of electrons is even or odd, respectively [138–140].

8.2 Wave-Particle Duality of Oil Spill

The position of an electromagnetic wave within the electromagnetic spectrum can be characterized by either its frequency of oscillation or its wavelength. Quantum electrodynamics is the theory of how EMR interacts with matter on an atomic level.

Carbon atoms cannot only behave like particles, but also like waves. This quantum-mechanical property is well-known for light particles such as electrons or hydrogen atoms. However, researchers have only rarely observed the wave-particle duality of heavy atoms, such as carbon. In other words, carbon atoms can display quantum effects, specifically, researchers observed that carbon atom can tunnel. The probability of an object being able to tunnel depends on its mass. The phenomenon can, for instance, be observed much more easily for light electrons than for relatively heavy carbon atoms. A hydrocarbon compound is identical to the product molecule. The same chemical compound thus exists before and after spilling into the ocean. However, the bonds between the carbon atoms change during the spreading on the ocean surface.

8.3 Quantum of Oil Spill Electric Conductivity

It is important to understand the electric conductivity of the oil spill to answer why the oil spill appears as dark patches in the SAR data. To answer this question another significant question arises, what is the quantum of crude oil spilled onto the sea surface? The answer to these questions leads to a new comprehensive approach for oil spill imaging by microwave remote sensing technology. The sea water consists of sodium chloride, which is dissolved in water. The sodium atom has a positively charged ion and the chlorine atom has a negatively charged ion. An ion is an atom or the cluster of atoms that have a negative or a positive electric charge. In this sense, the seawater contains an excess of ions than the crude oil [141–143]. The seawater ions will locate themselves around the crude oil particles to neutralize the surface charge. This accumulation of ions is termed as the electrical double layer, which consists of two layers—an inner layer and an outer layer. In this regard, the inner layer is also called the Stern layer, which is generated by ions of the opposite charge to the particle surface. These ions are known as counter ions and are adsorbed onto the particle surface [143]. On the contrary, the outer layer is a diffuse layer involving free ions that move under the influence of electrostatic attraction to the surface charge and comprises of both counter ions and co-ions—ions the of the same charge as the surface [144].

Consequently, the boundary locates between the inner and outer layers of the electrical double layer is known as the slip plane. In this view, the value of the electrostatic potential at this plane is termed as the zeta potential. In other words, the magnitude of the zeta potential bounces a sign of the potential stability of a colloidal system. In this understanding, when two charged surfaces of oil and seawater come close to each other, there will be electrostatic interactions between their electrical double layers. These approaches are a function of van der Waals attractions and electrostatic repulsion between the double layers, which describe the stability of crude oil and seawaters ion colloids [138]. These colloids are referred to as emulsions. In these circumstances, if the dispersed oil crude particles have a large negative, or a large positive, zeta potentials, they tend to repel each other and not fluctuate, thus there is higher colloidal stability. On the contrary, if the oil crude particles have low zeta potentials, the particles are not prevented by electrostatic repulsions and tend to flocculate. The dividing line of stable and unstable suspensions is usually taken at + 30 mV or –30 mV. However, the isoelectric point is achieved if zeta potentials equal zero. In this regard, the colloidal system is less stable. As a result, the zeta potential is a function of salinity. In this context, calcium ions can form 1:2 ion pairs with dissociated naphthenic acids, which can reduce the molecular level, interfacial tension (IFT) results from the difference in energy between molecules at a fluid interface compared to the corresponding bulk molecules. Interfacial tension is also correctly defined as a measure of how much mechanical energy is required to create a new unit area between two immiscible fluids. However, oil is a covalently-bonded substance, which is electrical insulators. In other words, a medium or substance with a dielectric property is an insulator [144].

8.4 Bragg Scattering and Dielectric Sea Surface

Sea surface has high dielectric which means high electrical conductivity. In this view, electrical conductivity is the capability of electrons to transfer through a sea surface. In quantum mechanics, the situation of an electron is regulated by its wavefunction $\Psi(x,t)$. Consequently, the motion of electrons is tackled from the point of view of the time evolution of the wave function, which is resolved by the Schrödinger equation:

$$i\hbar \frac{\partial}{\partial t} \Psi(\vec{x},t) = H\psi(\vec{x},t) \tag{8.7}$$

Equation 8.7 involves the Plank constant on its left side with the time derivative of the wavefunction, while the Hamiltonian H on its right side, which is given by Laplace operator Δ as:

$$H=-\frac{\hbar^2}{2m}\nabla^2 + E_p(x) \tag{8.9}$$

where m is the electron mass and E_p is the electron potential energy function of position x.

This potential energy of the electron relies on its surroundings and where it is positioned within that surrounding. In this sense, electrical conductivity is dealing with the motion of electrons through solids, for instance, metal or rubber. The atomic structure of most solids, conversely, is that of a crystal, which signifies that it has convinced regularities. For instance, the crystal structure of sodium chloride of seawater involves an ionic compound with the chemical formula NaCl (Fig. 8.2). In this understanding, the potential energy occurs by moving the electron through a crystal because the potential energy is just the sum of all of the interaction energies between the electron and the atoms [145–147]. In this regard, as the propagation of the electron through the crystal of NaCl would have similar symmetries as the crystal itself.

At the quantum mechanical level, electrical conduction amounts to a transition between electronic states. The conventional approach of this process, which is a transition from an electron being initially at one location, and then responding to an electromagnetic force and moving to another location. In quantum mechanics the probability of starting in one location and moving to another is considered, however, it turns out to be mathematically less convenient. An easier approach is to classify the electron positions consistent with their energies. An electronic state has certain energy when it is an eigenstate of the Hamiltonian operator [148,149]. That is, the corresponding wave function satisfies the time-independent Schrödinger equation,

$$H\psi(x)=E\psi(x) \tag{8.10}$$

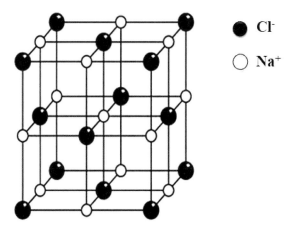

Figure 8.2. The ionic formula of sodium chloride.

where E is the energy, which is constant. Owing to the periodic potential energy function E_p, the allowed values of energy are assembled into bands (Fig. 8.3).

Figure 8.3 demonstrates that free electron energy depends on a wave vector $\mathbf{k}$. This energy can be indexed into number n, which is $n=1,2,....., N$. In this context, the free electron energy is defined as:

$$E = E_N(K) \tag{8.11}$$

The wave function of Equation 8.11 can be given by:

$$\Psi_{nK}(x,t) = e^{i\left[K.x-(E_{nK}/\hbar)t\right]}\varphi_{nK}(x) \tag{8.12}$$

Equation 8.12 takes the form of a Bloch wave, which is named after the Swiss physicist Felix Bloch, is a type of wave function for a particle in a periodically-repeating environment, most commonly an electron in a crystal (Fig. 8.4). φ_{nK} is a function with the same periodicity as the crystal. In addition, Equation 8.12 is considered as a solution to the full (time-dependent) Schrödinger equation. In this regard, it is similar to a plane wave propagating in the direction of $\mathbf{k}$, with frequency $\omega = E_{nK}\hbar^{-1}$ and a stationary envelope given by φ_{nK}. However, $\hbar K$ is not the real momentum of the electron, but is well-known as crystal momentum [145,149].

Since the Hamiltonian operator does not transform with the momentum operator in this state, having certain energy means that the electron does not have a definite momentum—it has a distribution of momenta. The average group velocity $\langle V_g \rangle$ of electrons through the crystal with energy E_{nK} is defined as:

$$\langle V_g \rangle = \hbar^{-1} \frac{\partial E_{nK}}{\partial K} \tag{8.13}$$

Equation 8.13 provides the gradient of the energy in K-space divided by the Planck constant $\hbar$. The average momentum $\langle P \rangle$, therefore, is formulated as:

$$\langle P \rangle = m\left[\hbar^{-1} \frac{\partial E_{nK}}{\partial K}\right] \tag{8.14}$$

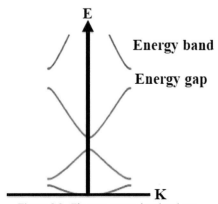

Figure 8.3. Electron energy band and gap.

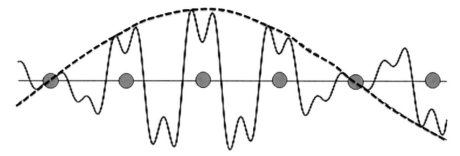

Figure 8.4. Bloch wave diagram.

The electrons in these energy bands begin to transfer around due to the application of an electric field to the solid. When the electric field slowly varies and is sufficiently small, the electrons move around only within the bands, changing between diverse quantities of K. For instance, when a constant electric field **E** is operated to the material, the wave vector of each electron after a (short*) time t is

$$K(t)=K(0)-\hbar^{-1}et\,E \tag{8.15}$$

where K(0) represents its initial wave vector and (−) e presents the electrical charge of the electron. A change of the wave vector, conversely, is compatible with a modification in the direction or velocity of the propagating wave function. Basically, the wave function of a separate electron accelerates as a complete in reaction to the electric field. On a large scale, accelerating electrons can be observed [142,144].

Nevertheless, if a band is totally occupied, there is nowhere in K-space for any of the electrons in that band to move without violating the Pauli exclusion principle. In this case, the electrons in this band remain stationary—they do not respond to the electric field. Hence, the oil spill particles where all the bands are either occupied or unfilled are an insulator: when a voltage is applied across it, there is little or no current generated. In contrast, the maximum filled bands within oil spill particles are only incompletely occupied. This causes plenty of space in those "conduction" bands for the electrons to propagate around, preceding to a more noticeable reaction to applied electric and magnetic fields. In this understanding, oil spill particles are not virtuous conductors [147,149]. However, when the electron acquires along the edge of the Brillouin zone in K-space of sea water particles, it reflects in a process similar to the Bragg reflection. This is why the electrons do not accelerate indeterminately, and a time-averaged stable current in reaction to the electric field is then established.

8.5 Impact of Surface Dielectric in SAR Backscatter

It is well known that the reflection and backscatter from the ocean surface rely on the wavelength distribution of the surface waves corresponds to the beam wavelength λ. In this case, the incident SAR beam is scattered from the capillary waves and reflected from long waves. In this view, the model of a unique reflection angle loses its significance in the circumstance of a wind-roughened the ocean surface. In fact, the ocean surface is turned only by wind-driven capillary waves. Therefore the wave surfaces can be approximated as a collection of congruent isosceles triangles, which are termed as facets, which serve as a specular reflector. Conversely, this approximation is valid if the length scale of each facet is much greater than λ and if the deviation of the approximating facet from the wave surface is much less than λ. Equivalently, the approximating holds if the radius of curvature R_c of that part of sea surface approximated by the facet satisfies:

$$R_c \gg \lambda \tag{8.16}$$

In this case, Equation 8.16 satisfies that, the beam fields at the sea surface can be approximated by the fields, which occurs a tangent plane. In other words, the composite of sea surface involves a large-scale surface, which satisfies the radius-of-curvature condition of Equation 8.16 and a small scale surface with a root mean square amplitude much less than λ. Therefore, the impact of the large scale surface is to advect and tilt the wind-induced patches of small scale roughness [150]. In this case, the wavelength separates the dual scales of the order of λ. Subsequently, this mechanism is a function of a complicated dependence on the root mean square amplitude (roughness) and incidence angle [151].

In the conventional concept, the specular reflection occurs because there is no wind and the ocean surface is flat. Since the wind speed increases and creates sea roughness the incoherent scatter increases, but coherent specular reflection decreases. If the incidence SAR beam is perpendicular to the wave-covered surface, the antenna receives specular reflection, but the incoherent beam is scattered away from the antenna (Fig. 8.5). In other words, the radar cross-section σ_0 decreases with increasing wind speed because of the maximum ocean wave slope rarely exceeds 15°. In this circumstance, σ_0 decreases as incidence angle $\theta < 15°$ under the increment in the wind speed. However, for $\theta < 15°$, the specular reflection is not received by an antenna (Fig. 8.6). However, incoherent backscatter only occurs in the antenna direction

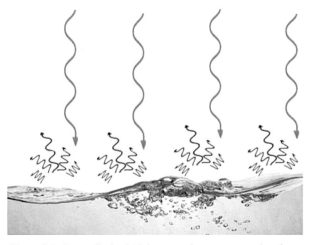

Figure 8.5. Perpendicular SAR beams to the wave-covered surface.

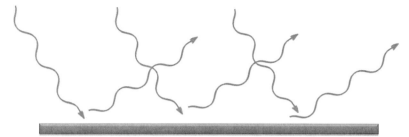

Figure 8.6. Specular reflection with $\theta < 15°$.

under the circumstance of oblique incidence SAR beams, i.e., $\theta < 15°$ (Fig. 8.7). In this regard, the specular reflection is mostly absent with the incidence angle as large as 70°, but strong oceanic backscatter is received by SAR antenna.

From the point view of quantum mechanisms, we say that specular reflection can cause such low electrical conductivity, which appears as the dark patches in SAR images. On the contrary, Bragg scattering can cause tremendous fluctuation of electron flows across the sea surface which generates a high electrical and by the way cause strong backscatter in SAR data. Generally, the Bragg scattering is delivered by the large angle backscatter. In this understanding, the specific incidence angles and frequencies, the backscatter exhibited strong resonance. For the ocean, if the surface wave spectrum involves a wavelength component with a similar relation to the incident beam, Bragg resonance occurs (Fig. 8.8). In other words, Bragg resonance occurs if there exists a surface component with λ_w equals to half the surface projection of the radar wavelength λ or when

$$\lambda_w = 0.5\lambda / \sin\theta \qquad (8.17)$$

Equation 8.17 indicates that the intensity backscatter from two contiguous wave crests is in phase. In this case, the SAR beams are incoherently backscattered from the sea surface enlarging coherently at the SAR antenna. In other words, the antenna received strong backscatter as a function of $\theta < 15°$. Generally, the wind creates a continuous spectrum of the capillary wave, which leads to a resonant wave. In this view, the mean–square wave slope and surface roughness increase with wind speed as well as Bragg scattering. In other words, Bragg scatters also occur from capillary wave riding on the long swell.

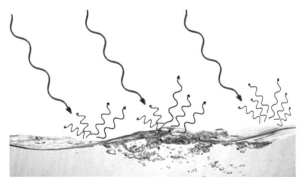

Figure 8.7. Incoherent backscatter with $\theta < 15°$.

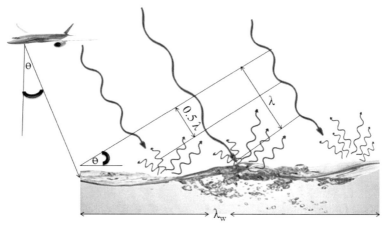

Figure 8.8. Bragg Scatter.

Consistent with the composite sea surface model, the roughness of the sea surface can be seen as small scale capillary waves superimposed on a large scale wind wave or swell. Conversely, the backscattered radar signal can be treated as Bragg scattering modulated by the tilted scattering surface which is created by large scale gravity waves. The mathematical expression of the total quantity of backscattered from the sea surface is formulated as:

$$\sigma_w = \sigma_s + \sigma_B \tag{8.18}$$

here σ_w is the total backscatter from the entire sea surface, which involves specular reflection σ_s and Bragg scattering σ_B. The damping of oil slicks on the short-gravity and capillary waves can reduce such specular reflection.

σ_c being the composite surface cross-section and the assured wave cross-section σ_b, which are expressed, respectively, by:

$$\sigma_c = \iint \sigma_B(\theta_o + \gamma, \, \alpha) \, P_f \, P(\gamma, \, \alpha|f) \, d\gamma d\alpha \tag{8.19}$$

$$\sigma_b = \iint \sigma_B(\theta_o + \gamma, \, \alpha) \, P_b P(\gamma, \, \alpha|b) \, d\gamma d\alpha \tag{8.20}$$

Both equations are based on standard Bragg scattering cross section σ_B as associated with three sorts of wave probabilities: (i) the probability of finding free waves P_f; (ii) the probability of finding bound waves (P_b); and (iii) the probability distribution of a wave sort, either free or bound ($P(\gamma, \, \alpha|x)$). Therefore, γ, α are the long wave slopes in and perpendicular to the plane of incidence and θ_o is the nominal incidence angle. Thus, on standard Bragg scattering cross section σ_B can be obtained by:

$$\sigma_B = 16 \, \pi \, k_o^{\,4} |F \, (\theta_o + \gamma, \, \alpha)|^2 \, \mu(2k_o \sin \, (\theta_o + \gamma), \, 0) \tag{8.21}$$

In Equation 8.21, the microwave number is k_o, the wave height variance spectrum is μ which is a function of $(2k_o \sin \, (\theta_o + \gamma), \, 0)$. Finally, F is a function of incidence angle θ and dielectric ε.

In general, the intensity of the radar cross section is a function of the incidence angle. If the radar signal strikes at right angles or at a high incidence angle to the surface, it will reflect more strongly than if it strikes at a low, grazing angle. In the latter case, much of the radar energy will be reflected away from the radar receiver, and it will, therefore, appear as a low or dark response on the image. Consequently, the local incidence angle is a major controlling factor in the intensity of the Bragg scattering cross section σ_B.

The radar backscatter tends to be the strongest from the slope of the wave facing towards the radar. Thus an image of backscatter modulated by tilt alone would represent a plane parallel swell-wave field as a series of parallel light and dark lines corresponding with slopes facing towards and away from the radar, i.e., displaced by 90° of phase from lines of troughs and crests.

In general, SAR ocean surface assumes rely on the very high sensitivity of incident angle and radar backscatter signal which are allied with fluctuations on both the local geometry and the spectral density distribution of short gravity and gravity-capillary waves. The resonant surface waves are much shorter at more oblique incidence angles. In other words, the ocean backscatter returns decrease with the increase of incidence angles. Consequently, the large oblique angles view smaller amplitude of Bragg waves leading to lower backscatter. Next the impact of the oil spill on sea surface roughness from the point view of quantum mechanisms, which can be detected from SAR data will be discussed.

8.6 Quantization Specular Reflection in SAR Data

Up to date, there is no novel explanation to address why the oil spill reduces the Bragg scattering in radar images. The theory is restricted to the conventional fashion of the dampening theory, which is addressed in the entire literature review of SAR imaging oil spills.

From the point view of quantum electrodynamics (QED), the oil spill particles covered in water do not allow the microwave photon to backscatter towards SAR antenna. In this view, the Rayleigh scattering processes with the oil spill particle cross-section is proportional to the fourth power of the angular photon frequency. Consequently, the oil spill particles induce the equivalent between incident microwave

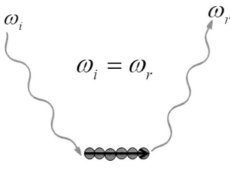

Oil Spill Particles

Figure 8.9. Sketch of specular reflection due to an oil spill from the point of view of QED.

photon frequency ω_i and the reflected photon due to oil spill particle, i.e., $\omega_i = \omega_r$ (Fig. 8.9).

Furthermore, a single microwave photon is simultaneously scattered away from the SAR antenna by all oil spill atoms forming a quantum superposition. The absent explanation is that the atoms of oil spills act as a bad conducting surface, which causes an efficient specular reflection. In this view, oil spill atoms have a partially empty electron energy band overlapping the valence band. In other words, oil spill atoms cause low energy specular reflection in SAR image with a frequency near to the atom's resonant frequency.

The physical properties that describe the specular reflection of microwave photon or the SAR signal by oil spill atoms can be deduced using only quantum electrodynamics. That is, the processes of emission and absorption of photons by atoms ultimately lead to the law of specular reflection and frequency invariance between the incident and specular signal reflection, and a quartic frequency dependence of the scattering cross section. Next quantum radar cross-section of oil spills will be explained clearly.

8.7 Quantum Radar Cross Section of Bragg Scattering of Oil-Covered

Consistent with the Bragg scattering theory, the Normalized Radar Cross Section (NRCS) is proportionate to the spectral energy density of the Bragg waves, i.e., of those surface waves of wavelengths λ_w that satisfy the Bragg resonance circumstance [152]. According to Gade and Alpers [153], the Normalized Radar Cross Section (NRCS) for a slick-free sea water surface (σ_w) can be formulated as:

$$\sigma_w = \left|g_{ij}\right|^2 \left(16\pi k_r^4 \cos^4 \theta\right) \psi\left(k_B\right) \tag{8.22}$$

where $\psi(k_B)$ is the spectral power density of the Bragg waves with the wavenumber $k_B = 2\pi\lambda_w^{-1}$, while k_r is the radar wavenumber. The function g_{ij} relies on radar wavenumber, the dielectric constant ε_w of the water, incidence angle and polarization. In thi regard, *ij* involves HH, VV HV, and VH; H and V mean horizontal and vertical polarization, respectively. The first letter denotes the polarization of the transmitted radiation and the second letter denotes that of the received radiation). The thickness of a surface film (either a monomolecular slick or a thin oil spill) is small compared with the penetration depth of microwaves into the water; therefore we may assume that the Bragg coefficient is not affected by the presence of a thin surface film [151]. Conversely, the ratio of the radar backscatter of horizontal and vertical polarization is independent of coverage of the water surface with a slick [152].

Consistent with the above perspective, SAR imaging mechanism of the oil spill is mainly a function of the radar cross-section gradient along the slick. Conversely, the radar-cross section from the point view quantum mechanism involves the photon wave function due to the interaction between the signal photons and atoms of the oil spill on the sea surface. In this regard, the signal photon wave function is formulated as:

$$\psi_\gamma\left(\Delta R, t\right) = \frac{\varepsilon_0}{\Delta r} \Theta\left(t - \frac{\Delta R}{c}\right) e^{-\left(0.5(i\omega+\Gamma)(t-\frac{\Delta R}{c})\right)} \tag{8.23}$$

Equation 8.23 involves Θ step function merely indicates that the signals will not propagate faster than the speed of light. Further,

$$\varepsilon_0 = -\frac{\omega^2 |\hat{\bar{\mu}}| \sin\eta}{4\pi\epsilon_0 c^2 \Delta r} \tag{8.23.1}$$

where ω is the frequency of the incoming signal photon, ϵ_0 is a polarization basis vector, either linear or circular, and η is the angle between the electric dipole moment of the atom μ. In addition, Δr is the distance between oil spill particles and SAR. Conversely, Γ is formulated as:

$$\Gamma \equiv \tau^{-1} = \frac{1}{4\pi} \frac{4\omega^3 |\hat{\bar{\mu}}|^2}{3\hbar c^3} \tag{8.23.2}$$

Equation 8.23.2 represents the inverse reflected signal photon delay time. In fact, the oil spill is considered in the dynamic system due to oscillation of the sea surface. The quantum of the intensity of signal illumination from the sea surface and oil slick covered is given by:

$$< \hat{I}_s(r_s, r_d, t) >= N^{-1} \left| \sum_{i=1}^{N} \psi_\gamma (\Delta R, t) \right|^2 \tag{8.24}$$

where ΔR is SAR range. Equation 8.24 describes that the SAR image is the fluctuations of the photon wave function which is reflected due to its interaction with an infinity of particles on the ground. From the quantum point view, the radar cross-section of oil spill covered and the clean sea is just a quantum radar cross-section which measures the rate of photon interaction with oil spill particles and ocean wave-particle under the circumstance of perturbation concept, i.e., Bragg scattering. In this case, radar cross-section in the form of quantum is σ_Q which measure the low-reflectivity of the oil spill covered. In other words, σ_Q is measures the quantum illumination of oil spill covered. In this sense, the quantum radar cross-section of the oil-covered is formulated as:

$$\sigma_Q = \lim_{R\to\infty} 4\pi R^2 \frac{< \hat{I}_s(r_s, r_d, t) >}{< \hat{I}_i(r_s, t) >} \tag{8.25}$$

where $< \hat{I}_i(r_s, t) >$ and $< \hat{I}_s(r_s, r_d, t) >$ are the incident and scattered energy density, respectively. Furthermore, r_s and r_d are the positions of the transmitter and the receiver. For quantum radar applications, Rayleigh scattering is the process that characterizes the interaction between the SAR signal photons and oil covered. That is, the final state of the atom is the same as the initial state, the circular frequency of the scattered photon is equal to that of the incident photon and the photon energy is low relative to the ionization energy of the interacting atom. The interaction between the photon and an atom is described using a fundamental QED process of absorption and emission of photons [154]. And the scattering effect only occurs in the two-order approximation of perturbation theory [155]. Let us assume the oil-covered area is $A_\perp$ proportional inversely with the power spectra of Bragg waves $\psi(k_B)$. In other words, as the oil spill covered area increases , the power spectra of the Bragg wave decreases. Consequently, the corresponding analytical expression of radar cross-section of oil spill covered can be obtained:

$$\sigma_Q = 4\pi \left(\frac{\sigma_w}{\left|g_{ij}\right|^2 \left(16\pi k_r^4 \cos^4 \theta\right)} \right) \lim_{R \to \infty} \frac{\left|\sum_{i=1}^{N} e^{\left(j\omega\Delta R_i c^{-2}\right)}\right|^2}{\int_0^{2\pi} \int_0^{0.5\pi} \left|\sum_{i=1}^{N} e^{\left(j\omega\Delta R_i' c^{-2}\right)}\right|^2 \sin\theta_d \, d\theta_d \, d\phi_d}$$

(8.26)

Equation 8.26 confirms that oil covered in a radar system is imagined area of the absent of Bragg wave power. In this view, the absence of Bragg wave in oil-covered is due to the collapsing of the photon wave function (Equation 8.26) due to the interaction with oil particles. Indeed, oil spills can drive across the sea surface in the form of wave-duality. This leads to the quantum decoherence when a superposition forms between the quantum signal photon states and the oil spill particles. In other words, the SAR mechanism for oil spill imagine theory is just considered as quantum decoherence, where the loss of information about Bragg waves from a system into the environment [156]. Consequently, decoherence ensues when diverse portions of the photons wave function develop entangled in different ways with the SAR receiver [157].

8.8 Decoherence Imaging Mechanism of Oil Spill

Coherence is the immobile relationship between waves in a beam of electromagnetic (EM) radiation. Two wave trains of EM radiation are coherent when they are in phase, in which they vibrate in harmony. In terms of the application to things like radar, the term coherence is also used to describe systems that preserve the phase of the received signal. In other words, in a pulse radar system, coherence describes the phase relationships between the transmitted and the received pulses (Fig. 8.10). Oscillations

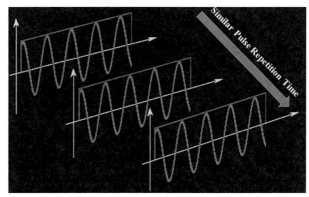

Figure 8.10. Radar coherence pulses.

and electromagnetic waves are described as coherent if their phase relationships are constant. In the case of incoherence, these phase shifts are statistically distributed. Whether a radar is coherent or no longer is decided through the kind of transmitter. As transmitters, extraordinary systems can work in the radar, which is either coherent, in part coherent or incoherent.

On the contrary, noncoherent occurs when a specified transmitter system is the self-oscillating Power Oscillator Transmitter. If such a transmitter is switched on and off by the almost right-angled high-voltage modulation pulse, this transmitter commences oscillating with a different phase shift with each transmitted pulse (Fig. 8.11). This phase shift, which occurs during the onset of the oscillation, is a virtually arbitrary process.

In the oil-covered pixels, the Bragg wave particles get washed out by interactions between the quantum particles and their surrounding environment. In this view, this interaction meant that oil spill particles and their surrounding Bragg wave particles become entangled: the properties of Bragg scattering or Bragg wave particles are no longer intrinsic to it but rely on the lowest dielectric of oil spill particles. This leads to the vanishing of rough Bragg wave intensity in the oil-covered water through SAR data. In the quantum view, the decoherence must be suppressed, by making the oil spill-covered pixels signed by dark spots to be isolated from the surrounding rough Bragg wave particles. In this regard, once the Bragg-wave or scattering washed away in SAR image by decoherence of oil spill pixels, the information of Bragg scattering covered by the oil spill cannot be obtained back. The rate of decoherence—the speed at which quantum superposition vanish-increases exponentially as the number of oil spill particles in the SAR image increases, so big oil-covered zones become conventional almost instantaneously. Needless to say, oil spill dark spots in SAR data are considered as quantum decoherence state. In fact, the oil spill induces the loss of the quantum particle behavior of Bragg wave owing to the interaction of the constituent oil spill particles with their surrounding environment.

Let us assume that SAR has a quantum system (for example a particle), which is usually designated by a wavefunction, which is a vector $|\psi\rangle$ from its Hilbert space H. Nonetheless, this description is not perfect when the system is entangled with another system. In this case, it is more appropriate to exploit the *density matrix* (or *density operator*). For $|\psi\rangle$, the density matrix is $\rho = |\psi\rangle\langle\psi|$ In general, it has the form

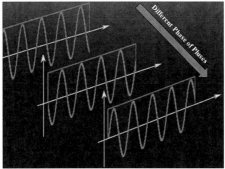

Figure 8.11. Noncoherent radar pulses.

$$\rho = p_i \sum_i |\psi_i\rangle\langle\psi_i| \tag{8.27}$$

Equation 8.27 is based on orthogonal states $|\psi_i\rangle$ from H. The basis can be selected contrarily, and this will offer other amounts for the coefficients p_i. Nevertheless, in general, a diverse selection would yield non-diagonal terms, that is, terms of the form $p_{ij}|\psi_i'\rangle\langle\psi_j'|, i \neq j$. The density matrix has some properties like being Hermitian (self-adjoint), positive semi-definite and has the trace $tr(\rho)=1$. In this regard, a density matrix which can be formulated as $\rho = |\psi\rangle\langle\psi|$ iff $\rho^2 = \rho$.

In this sense, oil spill caused by SAR image can be presented in the form of density matrix which involves a statistical ensemble of state vectors $|\psi\rangle$ as a function of probability p_i. In other words, let us assume, the observed dark spots in SAR images which belong to oil spill pixels have eigenstates the vector $|\psi_i\rangle$ so the detection of oil spill probability is p_i. On the contrary, look-alikes must have another basis of eigenstates the vector $|\psi'\rangle$. In this regard, the density matrix of look-alikes has non diagonal terms $p_{ij}|\psi_i'\rangle\langle\psi_j'|, i \neq j, p_{ij} \neq 0$. Here the probability can accurately judge which pixel in SAR image belong to the oil-covered water or look-alikes.

A significant question can be raised as: how can the density matrix of a SAR image assigned oil spill on the diagonal, and all of the other features being zero? This is achieved based on the decoherence as the density matrix of oil spills becomes diagonal with the sense to the foundation of eigenstates of the SAR data. As well as the oil spill pixels or look-alikes are in the diagonal form, it is easy to interpret as a statistical ensemble, having the probability p_i to acquire the I-th eigenvalue as the result of oil spill imaging and detection. In this view, quantum decoherence is then the procedure by which the density matrix develops to accurately ensure that the diagonal terms achieve oil spill occurrence, while the non diagonal terms achieve look-alike occurrences.

In general, we can formulate a simple rule to implement the quantum decoherence as a rule to explain the SAR mechanism for oil spill imagine. In this context, it can be suggested that decoherence is the phenomenon where a superposition of oil spill and the Bragg wave quantum states becomean incoherent mixture of those states in SAR image. Since both the oil spill and Bragg waves develop entangled with the SAR backscatter differently, which can be formulated as:

$$\left(|\psi_{Oil}\rangle + |\psi_{Bragg\ wave}\rangle\right)|SAR\rangle \tag{8.28}$$

To the state

$$|\psi_{Oil}, SAR\rangle + |\psi_{Bragg\ wave}, SAR\rangle + |\psi_{Look-alikes}, SAR\rangle \tag{8.29}$$

and since the environment states are different, this is functionally equivalent to just having either ψ_{Oil} as an index to an oil spill or $\psi_{Look-alikes}$ as an index to look-alikes with some probability, but not a superposition of the two. It might be the credulous belief occurring between the oil spill and Bragg waves or look-alikes and Bragg waves.

Texture and Quantum Entropy Algorithms for Oil Spill Detection in Synthetic Aperture Radar Images

9.1 Textures Based-SAR Backscattering from Oil-covered Water

The coherent nature of SAR provides intensification to pixel-to-pixel disparities in image intensity (or a related magnitude) that is designated by the diminishing statistics and which can be signified as a fading arbitrary variable. The spatial irregularity in the scattering characteristics of the pixels that create an object, delivers growth to an inherent vision texture, which can be embodied as a texture indiscriminate fluctuating in SAR data [158]. Therefore, the existence of speckle noise in satellite and airborne SAR images creates the images squat quality in standings of textural structures and spatial resolution, which are compulsory for detection of oil spill issues such as slick classification and clustering. In a manifestation of the prevailing speckle noise of SAR images, the oil spill and the surrounding sea environment cannot be efficiently detected in terms of radiometric (or pixel-based) and textural regions. Therefore, reduction of the level of speckle noise as the intrinsic or the inherent noise of SAR images assists to distinguish oil spill pixels from the surrounding environment.

The tone is formed in a SAR data by plug measurements of the backscattering coefficient (σ_0) and is essentially a function of the magnitude of interrelationships between the radar wavelength and the scattering components within a single pixel footprint. Texture denotes the spatial distinction of tonal elements as a function of scale [160]. Within a digital image, a homogeneous texture field consists of a spatial arrangement of gray levels which are more homogeneous (as a unit) within than between texture fields.

The SAR image data involves a backscattering signal fused with a coherent fading signal [158]. In fact, fading is the constructive and destructive intrusion of coherent electromagnetic radiation, which leads to SAR's speckles. In this view, speckle is visually accredited as an amplified frequency of light and dark pixels in what should be a relatively homogeneous gray level field. The statistics of coherent fading are well understood [158]. If we account for the fading component of image variance, the remaining image texture is primarily a function of the near surface characteristics. Therefore, image texture provides information which can be directly correlated with observable physical characteristics because different suites of these characteristics are associated with different sea surface features.

The texture component of σ_0 from the sea surface, which is a combination of surface roughness and volume scattering. Radar backscatter σ_0 is proportional to the roughness of the ocean surface, which is determined by the directional spectrum of the waves at the wavelength of Bragg resonance. For current spaceborne SAR systems, they are typically short gravity-capillary waves with a wavelength of a few centimeters. Oil spill on sea surface has a dampening effect on wind-generated short gravity-capillary waves, thus locally decreasing the radar backscattering. Consequently, oil film on the sea surface often corresponds to dark features in contrast to the brightness of the surrounding sea surface without an oil spill. Subsequently, identification of oil spill in SAR image has always comprised the first and an essential step detection of dark features.

Nonetheless, interpretation and detection of marine oil pollution features in the SAR data are regularly vague. In fact, other phenomena also show similar features of oil spills. These features are well-known as look-alikes. In this context, look-alikes appear as a dark footprint in SAR images which involve wind sheltering, current wakes caused by oil rigs, natural film, grease ice, low wind regions, rain cells, internal waves, current shear zones or natural seepages (Table 9.1). These features are considered as mere challenges for automatic detection of the oil spills. Because of the complexity

Table 9.1. Different characteristics of Oil Slick and Look-alikes in SAR Imagery.

Geophysical Expression	SAR image expression	Geophysical occurrence	Weather limitation	Backscatter value
Oil spill	Dark area	Not specific	3 m/s to 10 m/s	−25 to −10
Natural film	Reconfigures ease when interacting with current	Coastal regions and regions of upwelling	Dissolves at wind speeds above 7 m/s	−24 to −15
Wind sheltering by land	Dark regions near land	In the vicinity of land boundaries and in fjords	Even at high wind speeds (15 m/s)	−24 to −8
Threshold wind speed	Large dark areas	Not specific	Wind speed less than 3 m/s	−26 to −18
Rain cells	Bright cells with dark centers	Not specific	Heavy rain and strong winds	−24 to −8
Waves current interaction along shears	Narrow, bright or dark curving signatures	An area of strong currents	Wind speed less than 10 to 12 m/s	−24 to −8
Upwelling	Dark areas	Divergence zones of surface currents, mainly near the coast	Wind speed assumed less 6 to 8 m/s	−24 to −15

of the marine system and the relatively little information contained in SAR image, the automatic algorithm of oil spill identification in SAR image has not been well developed till now, and in most circumstances, oil spill detection nonetheless requests operators' aid [161]. Further system parameters such as frequency, polarization and incidence angle also play an important role in the amount of backscattered radiation that is received by the SAR antenna, and on the spatial arrangements of the gray levels (i.e., texture) in the SAR scene.

9.2 Texture Algorithms

Image texture analysis mainly contains three sorts, which are based on statistic, structure and spectrum. In this view, the texture analysis based on statistic computations is used to determine the binary characteristics of image textures. In this context, the binary texture characteristics can compile with non-texture characteristics, for instance, GLCM and Markov algorithms to classify the SAR image. Conversely, the texture analysis based on structure is used to find out the arrangement rules of texture primitive, such as mathematical morphology and topology. On the contrary, the texture analysis based on the spectrum is a function of multiscale analysis and time-frequency analysis, for instance, wavelet transform and fractal theory [159].

The texture is one of the important characteristics used to identify objects or regions of interest in an image. Unlike spectral features, which describe the average tonal variation in the various bands of an image, textural features contain information about the spatial distribution of tonal variations within a band. The Gray Level Co-occurrence Matrix, GLCM (also called the Gray Tone Spatial Dependency Matrix). The GLCM is a tabulation of how often different combinations of pixel brightness values (gray levels) occur in an image. The GLCM used in this book for a series of "second order" texture calculations. First order texture measures are statistics calculated from the original image values, like variance, and do not consider pixel neighbor relationships. Second order measures consider the relationship between groups of two (usually neighboring) pixels in the original image. Third and higher order textures (considering the relationships among three or more pixels) are theoretically possible, but not commonly implemented due to calculation time and interpretation difficulty. There has been some recent development of a more efficient way to calculate third-order textures.

9.3 Structure of the GLCM

9.3.1 Spatial Relationship between Binary Pixels

GLCM texture deliberates the relationship between binary pixels at a time domain, which is known as the reference and the neighbor pixel. The neighboring pixel is selected to be the one to the right of each reference pixel. This can also be articulated as a (1,0) relation: 1 pixel in the *x*-direction, 0 pixels in the *y*-direction. Therefore, every pixel within the kernel window turns out to be the reference pixel, which begins in the upper left corner of the SAR image and proceeds to the lower right. Pixels along the right edge have no right-hand neighbor, thus they are not depleted for this calculation (Fig. 9.1).

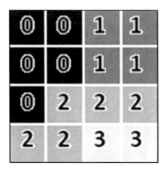

Figure 9.1. Reference and the neighbor pixels.

9.3.2 Separation Between Two Pixels

All the processed SAR data in this chapter applied a one-pixel offset, i.e., a reference pixel and its immediate neighbor. If the kernel window size is a large enough, training a larger offset is absolutely conceivable. There is no dissimilarity in the calculation technique. The sum of all the entries in the GLCM, i.e., the number of pixel combinations will just be smaller for a given kernel window size. Combinations of the gray levels that are conceivable for the trial image, and their position in the matrix. This is the only time that the labels in the top row and left column will be revealed. All further representations of the GLCM included only the 16 data cells. For instance, select the first pair of reference and neighbor pixels as indicated by the white circle in Fig. 9.2. Since the reference pixel has the gray level value of 0, and the neighbor pixel the value of 1, which can be counted as one entry in a table of frequencies (Fig. 9.3).

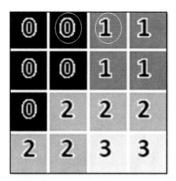

Figure 9.2. The separation between dual pixels.

9.3.3 Transposing

Transposing is creating the SAR matrix symmetrical around the diagonal. In this context, the texture calculations require a symmetrical matrix. The next step is,

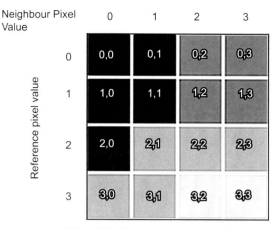

Figure 9.3. Frequency estimation table.

therefore, to get the GLCM into this form. Asymmetrical matrix means that the same values occur in cells on opposite sides of the diagonal. For instance, the value in cell 3,2 would be the same as the value in cell 2,3. The right matrix calculated above is not symmetrical.

The right matrix counted each reference pixel with the neighbor to its right. If counting is done this way, using one direction only, then the number of times the combination 2,3 occurs is not the same as the number of times the combination 3,2 occurs (for example 3 may be to the right of 2 three times, but to the left of 2 only once). Nonetheless, symmetry will be achieved if each pixel pair is counted twice: once "forwards" and once "backwards" (interchanging reference and neighbor pixels for the second count). A reference pixel of 3 and its eastern neighbor of 2 would contribute one count to the matrix element 3,2 and one count to the matrix element 2,3. Symmetry also means that when considering an eastern (1,0) relation, a western (−1,0) relation is also counted. This could now be called a "horizontal" matrix.

9.3.4 Normalization: The GLCM is Expressed as a Probability

After making the GLCM symmetrical, there is still one step to take before texture measures can be calculated. The measures require that each GLCM cell contains not a count, but rather a probability. Is a horizontal combination of, say (2,2), in the original image more likely than (2,3)? Looking at the horizontal GLCM shows that the combination 2,2 occurs 6 times out of the 24 horizontal combinations of pixels in the image (12 right + 12 left). In other words, 6 is the entry in the horizontal GLCM in the third column (headed reference pixel value 2) and third row (headed neighbor pixel value 2).

The simplest definition of the probability of a given outcome is "the number of times this outcome occurs, divided by the total number of possible outcomes." The combination (2,2) occurs 6 times out of 24, for a probability of 1/4 or 0.250. The probability of 2,3 is 1/24 or .042. Here is the equation to transform the GLCM into a close approximation of a probability table: It is only an approximation because a true

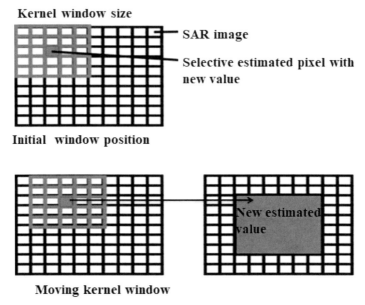

Figure 9.4. Texture estimation based on the kernel window size of 5 x 5 pixels and lines.

probability would require continuous values, and the gray levels are integer values, so they are discrete.

9.4 Creating Texture Image

Most texture calculations are weighted averages of the normalized GLCM cell contents. A weighted average multiplies each value to be used by a factor (a weight) before summing and dividing by the number of values. The weight is intended to express the relative importance of the value. For instance, the most common weighted average that students encounter is the term grade. Exams usually have a higher weight than quizzes. The weights are the % of course grade assigned to each mark. The result of a texture calculation is a single number representing the entire window. This number is put in the place of the center pixel of the window, then the window is moved one pixel and the process is repeated by calculating a new GLCM and a new texture measure. In this approach, an entire image is built up of texture values.

Each cell in a window must sit over an occupied image cell. This means that the center pixel of the window cannot be an edge pixel of the image. If a window has dimension $N \times N$, a strip $(N-1)/2$ pixels wide around the image will remain unoccupied. The usual way of handling this is to fill in these edge pixels with the nearest texture calculation. For a 5x5 window, the outer 2 rows and columns of the image receive the texture values calculated in row 3 (top edge), column 3 (left edge), row $L-2$ (bottom edge) and column $P-2$ (right edge) where P, L are the dimensions in pixels and lines of the original image. For the illustrated image, L=P=10, so values are calculated from row 3 and column 3 through row 8 and column 8.

Image edge pixels usually represent a very small fraction of total image pixels, so this is only a minor problem. However, if the image is very small or the window is very large, the image edge effect should be kept in mind when examining the texture image. Edge effects can be a problem in classification [162].

9.5 Mathematical Description of Co-occurrence Matrix

Co-occurrence is applied to categorize the image to an oil slick. Eight textures with 0°, 45°, 90°, 135° angular relationship and *d=1, 2, 3...., d_n* is employed such as energy; homogeneity; mean; variance; contrast dissimilarity; entropy; correlation. On the other hand, the SAR image gray tone can be described as texture (i.e., the microstructure). A co-occurrence matrix or co-occurrence distribution (less often co-occurrence matrix or co-occurrence distribution) is a matrix or distribution that is defined over an image to be the distribution of co-occurring values at a given offset. Mathematically, a co-occurrence matrix C is defined over an n x m image I, parameterized by an offset (Δ_x, Δ_y) as:

$$C_{\Delta_x, \Delta_{y(i,j)}} = \sum_{p=1}^{n} \sum_{q=1}^{m} \begin{cases} 1, & if \ x(p,q)=i and x(p+\Delta_x, q+\Delta_y)=j \\ 0, & Otherwise \end{cases} \tag{9.1}$$

The pioneer study of the implementation of texture algorithms was developed by Marghany [163], as the texture features used to discriminate between the oil spill and look-alikes.

9.5.1 Contrast group

It measures related to contrast use weights related to the distance from the GLCM diagonal. To emphasize a large amount of contrast, create weights so that the calculation results in a larger figure when there is a great contrast. Values on the GLCM diagonal show no contrast, and contrast increases away from the diagonal. So, create a weight that increases as the distance from the diagonal increases. It is also called the sum of squares variance:

$$Contrast = \sum_{i,j=0}^{N-1} P_{i,j}(i-j)^2 \tag{9.2}$$

when *i* and *j* are equal, the cell is on the diagonal and *(i–j)* = 0. These values represent pixels that are the same to their neighbor, so they are given a weight of 0. If *i* and *j* differ by 1, there is a small contrast, and the weight is 1. If *i* and *j* differ by 2, the contrast algorithm is increased and the weight is 4. The weights continue to increase exponentially as *(i-j)* increases. Since Contrast can evidently be < 1, it must be recorded in an image channel equipped to handle real numbers. If put into an 8-bit or 16-bit integer channel, the value would become 0 [164].

9.5.2 Dissimilarity

In the contrast measure, weights increase exponentially (0, 1, 4, 9, etc.), as one moves away from the diagonal. However, in the dissimilarity measure weights increase linearly (0, 1, 2,3, etc.).

$$Dissimilarity = \sum_{i,j=0}^{N-1} P_{i,j} \left| i - j \right| \tag{9.3}$$

Dissimilarity and Contrast result in larger numbers for more contrasting windows. If weights decrease away from the diagonal, the result will be larger for windows with little contrast.

Homogeneity is also called the "Inverse Difference Moment".

9.5.3 Homogeneity

Homogeneity weight values by the *inverse* of the Contrast weight, with weights decreasing exponentially away from the diagonal:

$$Homogeneity = \sum_{i,j=0}^{N-1} P_{i,j} \frac{P_{i,j}}{1 + (i-j)^2} \tag{9.4}$$

9.5.4 Angular Second Moment (ASM) and Energy

The name for ASM comes from physics and reflects the similar form of physics equations used to calculate the angular second moment, a measure of rotational acceleration. The meaning of "energy" is explained in the entropy link. The term Uniformity is perhaps less confusing. They are called uniformity. ASM and Energy use each P_{ij} as a weight for itself. High values of ASM or Energy occur when the window is very orderly.

$$ASM = \sum_{i,j=0}^{N-1} (P_{i,j})^2 \tag{9.5}$$

The square root of the ASM is sometimes used as a texture measure and is called Energy.

$$Energy = \sqrt{ASM} \tag{9.6}$$

Energy is, in this context, the opposite of entropy. Energy can be used to do useful work. In that sense it represents orderliness. This is why "Energy" is used for the texture that measures order in the image [165].

9.5.5 Entropy

It is a notoriously difficult term to understand; the concept comes from thermodynamics. It refers to the quantity of energy that is permanently lost to heat ("chaos") every time a reaction or a physical transformation occurs. Entropy cannot be recovered to do useful work. Because of this, the term is used in nontechnical speech to mean irremediable chaos or disorder. Also, as with ASM, the equation used to calculate physical entropy is very similar to the one used for the texture measure. Since $ln(0)$ is undefined, assume that $0 * ln(0) = 0$:

$$Entropy = \sum_{i,j=0}^{N-1} P_{i,j} (-\ln P_{i,j})$$
(9.7)

Entropy is usually classified as a first-degree measure, but should properly be a "zeroth" degree! The maximum value of *Entropy* is 0.5 [164,166].

For the GLCM Entropy algorithm:

i) P_{ij} is always between 0 and 1 because it is a probability and a probability can never be higher than 1 nor less than 0.

ii) Therefore, $ln(P_{ij})$ *will always be 0 or negative.*

iii) The smaller the value of P_{ij} (i.e., the less common is the occurrence of that pixel combination), the larger is the absolute value of $\ln(P_{ij})$.

iv) The (-1) multiplier in the entropy equation makes each term positive.

v)   Therefore, the smaller the P_{ij} value, the greater the weight.

For the mathematically inclined: More details about the maximum value of entropy from Lokesh Setia:

i)   The term $P * ln(P)$ is maximized where its derivative with respect to. P is 0. By the product rule, this derivative is $P * d(ln(P))/d(P) + d(P)/d(P) * ln(P)$ which simplifies to $1 + ln(P) = 0$, yielding $P = 1/e$. This means that the maximum of the term to be summed occurs when P is $1/e$, which is about 0.378.

ii) However, by definition the sum of $P_{ij} = 1$. With this constraint, the overall maximum of the sum (i.e., of ENT) is 0.5. This maximum is reached when all probabilities are equal. Conceptually this makes sense because when all probabilities of DN pairs are equal, we have a random distribution of DN values, which would yield maximum "chaos" or entropy [165–167].

9.5.6 Statistics of GLCM

The third group of GLCM texture quantities contains statistics derived from the GLC matrix. These are (i) GLCM Mean; (ii) GLCM Variance (or Standard Deviation) and (iii) GLCM Correlation. Similar statistics are derived from the GL values in the original image (not the GLCM) are also used as indicators of texture, but not of GLCM texture. For example, it is easy to calculate the image mean within a window. Nevertheless, the GLCM Mean is expressed in terms of the GLCM. The pixel value is not weighted by its frequency of occurrence by itself (as in a "regular" or familiar mean equation) but by the frequency of its occurrence in combination with a certain neighbor pixel value.

Because of this possible confusion, when any measure in this group is depleted, it should be clearly stated that it is the GLCM Mean, or Variance or Correlation that is being considered.

9.5.7 GLCM Mean

The mathematical description of GLCM mean is given by

$$\mu = \mu_i = \mu_j = \sum_i i \sum_j P(i,j) = \sum_j j \sum_i P(i,j) \qquad (9.8)$$

The mean is estimated based on the reference pixels, μ_i. It is also possible to calculate the mean using the neighbor pixels, μ_j, as in the right-hand equation. For the symmetrical GLCM, where each pixel in the window is counted once as a reference and once as a neighbor, the two values are identical.

The mean calculation is done as follows:

i) The summation is from 0 to (N-1), not from 1 to N. Since the first cell in the upper left of the GLCM is numbered (0,0), then the *i* value (0) of this cell is the same as the value of the reference pixel (0). Similarly, the second cell down from the top has an *i* value of 1, and a reference pixel value of 1. If this is not clear, go back and look at the framework GLCM.

ii) The P_{ij} value is the probability value from the GLCM, i.e., how many times that reference value occurs in a specific combination with a neighbor pixel. It is not a measure of how many times the reference pixel occurs time, would be the "regular" mean for the original window.

iii) Multiplying *i* by P_{ij} effectively divides the entry *i* by the sum of the entries in the GLCM, which is the number of combinations in the original window. This is the same as is done when calculating a mean in the "usual" way. If this is not clear, review how P_{ij} is calculated in the first place.

iv) The GLCM Mean for the horizontal GLCM is different from that for the vertical GLCM because the combinations of pixels are different in the two cases [165–167].

9.6 Can GLCM Accurately Detect Oil Spill?

GLCM different algorithms are implemented to two single look-complex RADARSAT-1 SAR data, which were acquired on October 26, 1997, and December 20, 1999, respectively along with the coastal water of the Malacca Straits. The wide mode 1 RADARSAT-1 SAR data with C-band (5.6 cm wavelength and frequency 5.3 GHz) with HH polarization was acquired in 1997 shows a wide area of dark spots which are parallel to the coastal waters (Fig. 9.5). In fact, on October 17, 1997, the collision in the Singapore Straits occurred between two ships, i.e, MT Orapin Global and MV Evoikos [163], and spilled about 28 000 tonnes of crude oil into the Malacca Straits.

Moreover, the second RADARSAT-1 SAR image was acquired on December 20, 1999 (Fig. 9.6) with Standard-2 beam mode, which has 12.5 m spatial resolution and 110 km x 100 km swath area with an incidence angle between 23.7°–31.0° (Table 9.2).

It is preferred to implement the GLCM to SAR data (7.9a) without any prior filtering (Fig. 9.7b). In fact, this procedure can cause the losing of image spatial

Figure 9.5. RADARSAT-1 SAR wide mode acquired on October 26, 1997.

Figure 9.6. RADARSAT-1 standard 2 beam mode acquired on December 20, 1999.

Table 9.2. RADARSAT-1 Modes characteristics used in this chapter.

Mode	Resolution (R[1] x A, m)	Look[2]	Width (km)	Incidence (°)
Standard	25 X 38	4	100	20–49
Wide (1)	48–30 X 28	4	165	20–31

[1]Nominal; ground resolution varies with range
[2]Nominal; range and processor-dependent

(a)

(b)

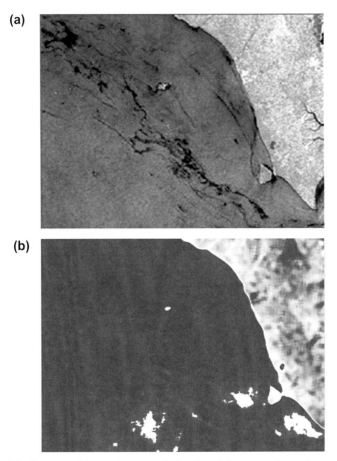

Figure 9.7. The result of (a) raw RADARSAT-1 SAR and (b) average filter prior GLMC analysis.

information (Fig. 9.7b). Consistent with Marghany [163] any filters applied to SAR image prior GLMC averages out a lot of texture in an image and leads to output result with a degraded image (Fig. 9.7b).

The contrast algorithm of GLMC performed on SAR image with the selected kernel window size of 3 x 3 pixels and lines (Fig. 9.8). In contrast to the unprocessed SAR image, contrast, correlation, second, dissimilarity, the energy produced poor spatial information in SAR image. These algorithms are not able to determine any spatial information about both oil spill and look-alikes. In the words of Conners and Harlow [168], contrast measures the amount of local variation in the image. It is high when the local region has high contrast in the scale of spatial variation. Contrast is a function of the large size of the oil spill (Fig. 9.8) because the contrast reflects both the coarseness of the texture and the contrast of the edge. This explains why the large size oil spill is detected in contrast images with a sharp shoreline edge.

The correlation image, therefore, mapped the area of dark spots with bright signatures (Fig. 9.8). In fact, the correlation measures the linear dependency of gray

Contrast   **Correlation**

Dissimilarity   **Second-moment**

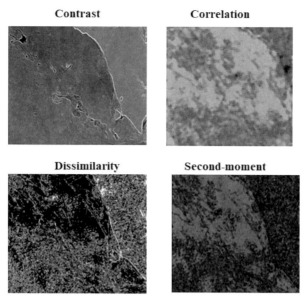

Figure 9.8. GLMC results of contrast, correlation, dissimilarity, and second-moment.

Original RADARST-1 SAR   **Mean Algorithm**

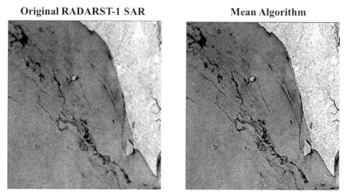

Figure 9.9. GLMC based Mean algorithm.

Color version at the end of the book

levels of neighboring pixels. As stated by Haralick [160], when the scale of local texture is much larger than the distance between pixels, correlation is typically high. When the local texture has a scale similar to or smaller than the distance between pixels, the correlation between pairs of pixels will be low. This explains why only the largest oil spill area is detected. However, the Mean algorithm seems to produce a despeckled image (Fig. 9.9). Indeed, Mean is expressed in terms of the GLCM. The pixel value is not weighted by its frequency of occurrence by itself (as in a "regular" or familiar mean equation) but by the frequency of its occurrence in combination with a certain neighbor pixel value.

On the contrary, Entropy (Fig. 9.10) performs better discrimination for an oil spill from the surrounding sea than Energy (Fig. 9.11) and Homogeneity (Fig. 9.12).

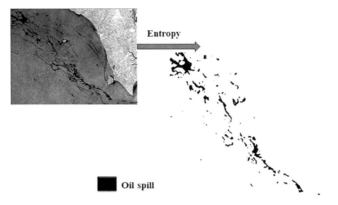

Entropy

Oil spill

Figure 9.10. GLMC based on entropy kernel window size 7 x 7 pixels and lines.

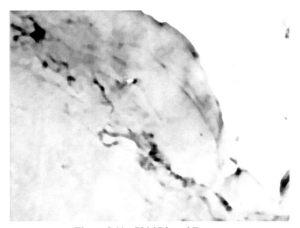

Figure 9.11. GLMC based Energy.

Figure 9.12. GLMC based on Homogeneity.

In fact, energy measures the non-uniformity in the image. This could be used to distinguish between the low backscattered value of oil spills and the surrounding pixels. Conversely, entropy is a measure of uniformity in the image, it can be used to distinguish between oil spill pixels and the surrounding environment.

In other words, Heterogeneity within the nodule was assessed with entropy and homogeneity. Entropy is a term that has been widely used in information theory as

Figure 9.13. Automatic detection of the oil spill by Entropy of RADARSAT-1 SAR image was acquired on December 20, 1999.

a parameter that reflects the unpredictability or information content of an image. Homogeneity is the measure that increases with less contrast in the kernel window and was calculated using the two-dimensional image histogram in this study. High entropy and low homogeneity values indicate increased heterogeneity of the nodule. In addition, both Energy and Homogeneity cannot distinguish between the oil spill and look-alikes. In this view Energy and Homogeneity just detect the intensity of dark pixels without isolating look-alikes from oil spill pixels as it is performed by Entropy.

In general, Entropy is able to isolate the oil spill from the surrounding environment. In this regard, another RADARSAT-1 SAR image was acquired on December 20, 1999, with Standard 2 (S2) beam mode confirms the performance of Entropy (Fig. 9.13). This also confirms the work of Marghany [163] that the kernel window size of 7 x 7 pixels and lines is the most appropriate to produce an automatic detection of oil spill using GLMC based of Entropy.

9.7 Can Quantum Entropy Perform better than Entropy for the Automatic Detection of Oil Spill?

The von Neumann entropy is the analog of the Boltzmann entropy in quantum mechanics. Any density matrix ρ can be written as

$$\rho = \sum_i p_i |i\rangle\langle i| \tag{9.9}$$

where p_i presents the probability of the state i which is considered as a probability distribution on state vectors. Therefore, the quantum entropies have a classical analog, which is mathematically formulated as:

$$\langle S \rangle = -K_B T_r (\hat{\rho} \ln \hat{\rho}) \tag{9.10}$$

where T_r is the trace so Equation 9.10 is the quantum version of the Gibbs Entropy which is given as:

$$\langle S_{cl} \rangle = -K_B \int dpdx(\rho \ln \rho) \tag{9.11}$$

For a statistical mechanical system with microstates 1,2,....,n and a probability p_i that the system will access microstate i in its evolution. Now, consider an n-dimensional quantum system with Hilbert space $\mathscr{H}$ and density operator ρ and Hamiltonian H. In a system in equilibrium, the density operator commutes with H:

$$[\rho, H] = 0 \tag{9.12}$$

let $|1\rangle, ...,|2\rangle$ denotes an orthonormal basis for $\mathscr{H}$, which consists of simultaneous eigenvectors of ρ and $\mathscr{H}$. On this basis, the density operator will be diagonal as:

$$\rho = diag(p_1, ..., p_n) \tag{9.13}$$

Moreover, because the density matrix is non-negative and self-adjoint, each diagonal entry p_i is non-negative and real, so the sequence $p_1, p_2, ..., p_n$ can be viewed as a probability distribution, where p_i represents the ensemble-probability that the system is in state |i⟩. Therefore, the difference between these two quantities of oil spills and look-alikes or the sea surrounding surface can be formulated as:

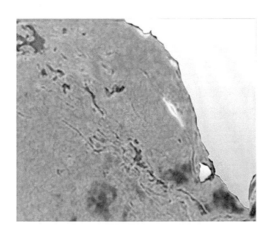

Oil Spill   Look-alikes   Sea surface

Figure 9.14. Quantum entropy for classification of sea surface features from RADARSAT-1 SAR wide mode acquired on October 26, 1997.

Color version at the end of the book

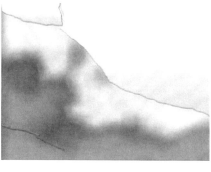

Oil Spill   Look-alikes   Sea surface

Figure 9.15. Quantum entropy for classification of sea surface features from RADARSAT-1 SAR acquired on December 20, 1999.

Color version at the end of the book

$$-\sum_j p_i \log q_j - \left(-\sum_j p_i \log p_j \right) = \sum_j p_j \log p_j - \sum_j p_j \log q_j \qquad (9.14)$$

Equation 9.14 is a measure of the distinguishability of the two probability distributions p and q for the oil spill and look-alikes, respectively. Figures 9.14 and 9.15 confirm that the quantum entropy algorithm is able to distinguish the differences in oil spills, look-alikes and sea surface characteristics. Entropy is only capable to detect oil spills automatically from its surrounding environments. However, the quantum entropy is proficient to perform such classification for the oil spill, look-alikes and sea surface pixels.

In quantum information theory, quantum relative entropy is a measure of distinguishability between two quantum states. It is the quantum mechanical analog of relative entropy. In fact, quantum entropy can measure the decoherence of oil spill particles when they interact with the rest of the surrounding environment. Moreover, density matrix involves in the quantum entropy allows for the description of a quantum system in a mixed state, i.e., coherence and decoherence. In this understanding, the quantum entropy interprets the SAR images in $\mathcal{H}$ spaces. In other words, SAR data can be interpreted as the space of the three different patterns, oil spills, look-alikes and the rough sea surface of one physical system of SAR image. In this case, the SAR image can mathematically be expressed as:

$$\mathcal{H} = \mathcal{H}^{\text{ oil spill}} \otimes \mathcal{H}^{\text{ look-alikes}} \otimes \mathcal{H}^{\text{ Sea surface}} \qquad (9.15)$$

From the point view of Equation 9.15 the density matrix $\rho^{oil \text{ look-alikes}}$ on $\mathcal{H}^{\text{ oil spill}}$ $\otimes$ $\mathcal{H}^{\text{ look-alikes}}$ as partial trace $\rho^{oil \text{ look-alikes}} = T_{r \, \mathcal{H} \, \text{Sea surface}} \rho^{oil \text{ look-alikes sea surface}}$, which proved by the output results in Figs. 9.14 and 9.15. Basically, what quantum entropy does to characterize the likelihood of a given class of outcomes based on the difference of quantum states.

Indeed, an oil spill surrounding environment generally induces some "uncertainty" in the spin direction which is responsible for quantum decoherence, i.e., for the rapidly vanishing of the off-diagonal elements of the spin reduced density matrix. On the other hand, a finite coupling between the spin of oil spill particles or Bragg wave particles and its environment also generates entanglement between the spin and the environmental bosons, i.e., the wavefunction of the system cannot be written as a simple product state anymore. In this view, it believes a great connection between decoherence of the oil spill particle spins and (entanglement) entropy, which may be seen as follows: the decoherence irreversibly converts the averaged or environmentally traced over spin density matrix from a pure state (Bragg wave) to a reduced mixture.

It is confirmed that the loss of coherence of the dual-level of Bragg wave and oil spill systems is characterized either by the complete destruction of Bragg oscillations or by the strong suppression of the persistent low dielectric of the oil spill in a SAR image. This means a prominent enhancement in the entanglement of the SAR image. In fact, this enhancement of entanglement corresponds to the actual quantum phase transition in the case of a rough Bragg scattering and rather corresponds to the dynamical incoherent crossover from damped oscillatory to overdamped by a low dielectric of oil spill particles. In fact, the decoherence occurs along the oil spill pixels due to the phase transitions from Bragg scattering to specular scattering [169].

Mahalanobis Classifier and Neural Network Algorithms For Oil Spill Detection

Machine learning is a keystone to construct automated systems, which can classify and recognize complex pattern spatial variations in such SAR data. It is not surprising therefore, that the demonstration of the SAR data as a coherent system provides a vital role in determining what kinds of patterns can be automatically revealed. In the previous chapters, the new concept of decoherence of oil spill pixels were addressed. The significant question is how machine learning algorithms such as Mahalanobis and neural network can be implemented as an automated classifier for the oil spill and look-a-likes in SAR data? In fact, there are few studies mentioned in oil spill in literature, which is implementing machine learning. The pioneer of learning machine studies for automatic detection of the oil spill is Konstantinos Topouzelis.

10.1 Machine Learning Algorithms for Automatic Detection of Oil Spill

There are dual techniques in which to handle the oil-spill detection to handle the complexity of the oil spill in SAR data: (i) grinding the characterization of the slick by using of the multi-polarization features of SAR techniques [5–7], and (ii) investigation of the brightness SAR backscatter signal without deliberating the constraints of the procedures of image formation and acquisition [170]. Most of the image processing techniques are based on segmenting the dark spots in the SAR data to discriminate between the oil spill and look-alikes. In this context, implementing statistical classifiers can be useful to distinguish between the oil spill and its surrounding environment [176,178].

    The foremost step in oil spill detection, therefore, is achieved by trained operators and entails visual scrutiny techniques and analysis of the extracting features from SAR images. Nonetheless, Topouzelis [170] stated that owing to the efficiency of machine learning-based techniques on remote sensing, semi-automatic or fully automatic approaches are the state-of-the-art in oil spill detection. In this regard, neural networks have been generally explored and spotted as a vital device for classification. Most of these automatic techniques are related to conventional Multilayer Perceptron (MLP) neural networks, probabilistic techniques and fuzzy classification, by means of considerable data sets for training and validation. In this view, Del Frate et al. [171] developed a semi-automatic detection of oil-spill by using the neural network of which the input parameter is a vector describing features of an oil-spill candidate. They proved that the neural network could accurately distinguish the oil-spill and look-alikes in SAR data. Topouzelis et al. [172] developed a new MLP neural network approach with one hidden layer (51 neurons), 10 feature input vector size and 2 output nodes. In this context, the MLP neural network system was trained and verified using 24 high resolution SAR images containing 69 oil spills and 90 look-alikes. NN topology was configured using a genetic algorithm. The accuracy reported on the test data was: 91% for oil spills and 87% for look-alikes. Further, Topouzelis et al. [173] also carried out a detailed robustness examination of the combinations derived from 25 commonly used features. They initiated that a combination of 10 features yields were the most accurate results. Garcia-Pineda et al. [174], therefore, developed the Textural Classifier Neural Network Algorithm (TCNNA) to map oil-spill in the Gulf of Mexico Deepwater Horizon accident by combining Envisat ASAR data and wind model outputs (CMOD5) using a combination of two neural networks. However, the ANN techniques are dominated by several disadvantages of time-consuming, and the high rate of the misclassification probability as the number of features increases. This leads to the matter of the ordeal of dimensionality, which also rises due to insufficient sample data. To overcome such a problem, Li et al. [175], developed a Support Vector Machine (SVM) to automatically detect oil-spill from SAR images. In this regard, a binary-threshold segmentation was implemented on SAR backscatter images to extract information in different grayscale levels. In the classification phase, a linear SVM was firstly built according to the training data set. Then the constructed SVM was used to distinguish probable oil slicks and look-alikes based on morphological characteristics extracted from the dilated low threshold segmentation result. The final detection result was obtained by fusing the classification result of both low and high threshold segmentation. In the comparison between Characteristic Possibility Function (CPF) based method, the SVM method has better performance in reducing both type I (incorrect rejection) and type II (incorrect acceptance) error. In comparison with ANN based methods, lower false alarm rate was obtained. Conversely, a comparison in the span of classification precisions between the cited classifiers would not be consistent owing to the usage of dissimilar SAR datasets, which are not continuously accessible. Moreover, the capricious quantity of the extracted features is reliant on the acquisition sensor, along with numerous classifier configurations.

Infact, numerous algorithms for machine learning designate that the SAR data are signified as components in a metric space. For instance, in prevalent algorithms such as nearest-neighbor classification, vector quantization and kernel density estimation, the metric distances between diverse features deliver a quantity of their dissimilarity. The routine of these algorithms can rely perceptively on the manner in which distances are measured, for instance, the Mahalanobis algorithm. In this understanding, when

SAR data are embodied as points in a multidimensional vector space, simple Euclidean distances are frequently implemented to quantify the distinction between oil spill pixels and their surrounding environment. Nonetheless, such distances frequently do not yield consistent decisions; as well as, they cannot emphasize the distinctive features that play a role in certain types of classification, but not others. Physically, it is required at different metrics for computing distances between feature vectors.

10.2 Hypotheses

This chapter proposes some novel machine learning algorithms automatic detection of the oil spill from SAR data. This chapter has also hypothesized that the dark spot areas (oil slick or look-alike pixels) and its surrounding backscattered environmental signals in the SAR data can be detected using post-classification techniques such as Mahalanobis classifier, and artificial intelligence techniques. The contribution of this work concerns the comparison between Mahalanobis classifier and artificial intelligence techniques. In fact, previous works have implemented post classification techniques [176–180] or artificial neural network [171–175] without comparing to other techniques. The main objective of this work can be divided into two sub-objectives: (i) To examine various algorithms such as Mahalanobis classifier; and artificial intelligence techniques [172] for oil spill automatic detection in multimode RADRASAT-1 SAR data; and (ii) To determine an appropriate algorithm for oil spill automatic detection in multimode RADRASAT-1 SAR data that are based on algorithm's accuracy.

10.3 Selected SAR Data Acquisition

The SAR data acquired in this study are from the RADARSAT-1 SAR that involves Standard beam mode (S2); W1beam mode (F1) image (Table 1). SAR data are C-band and have a lower signal-to-noise ratio due to their HH polarization with a wavelength of 5.6 cm and a frequency of 5.3 GHz [7–13]. Further, RADARSAT-SAR data have 3.1 looks and cover an incidence angle of 23.7° and 31.0° [12]. In addition, RADARSAT-SAR data cover a swath width of 100 km. According to Marghany [8], Marghany and Mazlan [13] oil spill occurred on 17 December 1999, along with the coastal water of Malacca Straits [16,18].

Table 10.1. RADARSAT-1 SAR Satellite Data Acquisitions.

Mode type	Resolution (m)		Incident angle (°)	Looks	Swath width (km)	Date
	Range	Azimuth				
W1	182.44	150.000	20°–5°	1 x 4	100 km	1997/10/26
S2	111.66	110.037	20°–41°	1 x 4	100 km	1999/12/17
F1	113.68	137.675	37°–49°	1 x 1	50 km	2003/12/11

10.4 Mahalanobis Algorithm

The MD was founded by a famous Indian statistician, P. C. Mahalanobis around 1936. The MD is said to be a distance measure based on correlations between the variables and by which different patterns could be identified and analyzed with respect to a base or a reference point. In this regard, let us assume that the Mahalanobis distance is a measure of the distance between a point P and a distribution D, which is a multidimensional generalization of the concept of measuring how many well-known deviations away P is from the mean of D. This distance is zero if P is at the mean of D, and grows as P strikes away from the mean alongside every major element axis. The Mahalanobis distance measures the number of trendy deviations from P to the mean of D. If each of these axes is re-scaled to have unit variance, then the Mahalanobis distance corresponds to the well-known Euclidean distance in the changed space. The Mahalanobis distance is as a consequence is unitless and scale-invariant and takes into account the correlations of the facts set.

In this sense, M.D. is an extremely useful approach of determining the "similarity" of a set of values from an "unknown", i.e., oil spill and look-alike pixels to a set of values measured from a collection of "known" SAR image. It is superior to the Euclidean distance because it takes a distribution of the points (correlations) into account. Conventionally transforms to classify clarifications into different groups, for instance, oil spill, look-alikes and sea surface. Figure 10.1 explains what the MD is; x, r, y, p, z, and w are expressing and visualizing the multidimensional space (system). The black arrow is the vector at a given point (the black dot as a reference point). The gray arrow is the vector to a random set point (the gray dot). The dotted arrow symbolizes the MD, or in relation to ecological analyzes the distance to the ecological signature.

The scientific explanation of Mahalanobis distance of a multivariate vector can mathematically be written as:

$$D_{M_{(v)}} = \sqrt{(v - \mu)^T C^{-1} (v - \mu)}.$$

(10.1)

where $V = (v_1, v_2, v_3 \dots \dots, v_n)^t$ from the group of values with mean $\mu = (\mu_1, \mu_2, \mu_3 \dots \dots, \mu_n)^t$, and S, is the covariance matrix. In order to apply the Mahalanobis

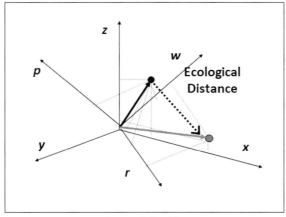

Figure 10.1. Mahalanobis concept.

classification procedures to different SAR data, let assume v is the feature vector for the unknown input, and $\mathbf{M}_1$, and $\mathbf{M}_2$ are the two classes: oil spill and look-alike. Then the error in matching v against $\mathbf{M}_j$ is given by the Euclidean distance $[v–\mathbf{M}_j]$. A minimum-error classifier computes $[v–\mathbf{m}_j]$ for j=1 to 2 and chooses the class based on the minimum error (Fig. 10.2).

If the covariance matrix is the identity matrix, the Mahalanobis distance reduces to the Euclidean distance. If the covariance matrix is diagonal, then the resulting distance measure is called the normalized Euclidean distance:

$$d = (\vec{v})\sqrt{\sum_{i=1}^{n} \frac{v_i^2 - v_j^2}{\delta_i^2}}. \tag{10.2}$$

where δ_i is the standard deviation of the v_i over the sample set. The input parameters from SAR data can impact the Mahalanobis classifier running when the variability of inner classes is smaller than the entire classifier group variance. In this context, if the class M is badly scaled and the decision boundaries between classes are curved, the classifier accuracy is reduced. Some of the limitations of simple minimum-Euclidean distance classifiers can be overcome by using the Mahalanobis distance d_t^2 that in the covariance matrix C form is

$$d_t^2 = (v - M_j)c^{-1}(v - M_j) \quad \text{pecularity.} \tag{10.3}$$

The Mahalanobis distance accounts explicitly for the different scales and correlations amongst vector entries and can be more useful in cases where these are significant.

The most important step involved in Mahalanobis algorithm is learning vector quantization. In this view, it uses a set of codebook vectors and seeks for the minimum of distances of an unknown vector to these codebook vectors as the criterion for

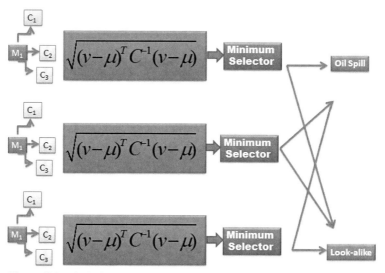

Figure 10.2. The logical operation of detection oil spill using Mahalanobis Classifier.

classification. However, unlike in conventional Mahalanobis algorithm where codebook vectors are fixed, an iterative procedure is adopted in which the position of the codebook vectors is adjusted to minimize the number of misclassifications. In this understanding, Optimized-Learning-Rate is used to perform learning of the optimal codebook vector positions. This algorithm has an individual learning rate for each codebook vector, resulting in faster training.

In this case, let assume a set of initial codebook vectors M_i ($i = 1, ..., k$) which have been linked to each class region, the input vector v is assigned to the class which the nearest M_i belongs, i.e., an NNC task is performed. Let c define the index of the nearest codebook vector, i.e., M_c. Learning is an iterative procedure in which the position of the codebook vectors is adjusted to minimize the number of misclassifications. At iteration step t let $v(t)$ and $M_i(t)$ be the input vector and codebook vectors respectively. The $M_i(t)$ is adjusted according to the following learning rule:

$$M_c(t+1) = \left[1 - s(t)\alpha_c(t)\right]M_c(t) + s(t)\alpha_c(t)v(t)$$
$$M_i(t+1) = M_i(t) \quad \text{for } i \neq c$$

(10.4)

where $s(t)$ equals $+1$ or -1 if $v(t)$ has been respectively classified correctly or incorrectly, and $\alpha_c(t)$ is the variable learning rate for codebook vector M_c:

$$\alpha_c(t) = \frac{\alpha_c(t-1)}{1 + s(t)\alpha_c(t-1)}$$

(10.5)

The learning rate must be constrained such that $\alpha_c(t) < 1$. The MD has been chosen as a multivariate index of the oil spill detection because of these advantages; it takes into account not only the average value but also its variance and the covariance of the variables measured. It accounts for the ranges of the acceptability (variance) between variables. It compensates for interactions (covariance) between variables. It is dimensionless. And if the variables are normally distributed, they can be converted to probabilities using the chi-square density function.

10.5 Oil Spill Detection by the Mahalanobis Classifier

The dark spots in the RADARSAT-1 SAR images are examined using the Mahalanobis classifier. Figure 10.3a shows large continuous dark patches along the coastline of the Malacca Straits. In the previous chapter, the entropy algorithm was able to identify these dark patches as an oil spill. Conversely, Mahalanobis classifier can identify two classes, which belong to the dark spots. It is interesting that the oil spill pixels are classified as dark patches, while look-alike pixels are identified as bright dark patches (Fig. 10.3b). Nonetheless, the standard beam mode (S2) of RADARSAT-1 SAR images does not indicate any dark patches that covered the sea surface. In this case, the Mahalanobis classifier does not classify either oil spill pixels nor look-alike pixels. In this view, it identifies two classes, sea surface and land (Fig. 10.4b).

The fine mode data F1 shows a wide dark patch close to the coastline (Fig. 10.5a). The Mahalanobis classifier is able to identify these dark patches as a thick oil spill,

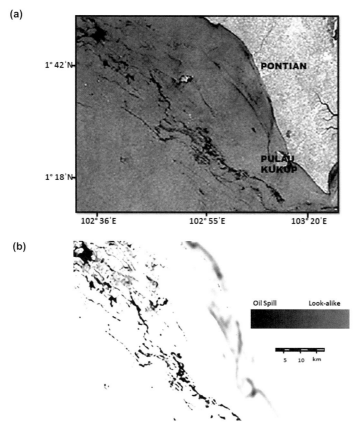

Figure 10.3. Examined RADARSAT-1 SAR (a) W1 mode and (b) Mahalanobis classifier.

thin oil spill, low wind-zone and look-alikes (Fig. 10.5b). This classification uses the training data by estimating the means and variances of the classes, which are used to estimate the probabilities and also consider the variability of brightness values in each class.

It is the most powerful classification method when accurate training data is provided and one of the major widely used algorithms [177]. For instance, Figs. 10.3 and 10.5 show that the slick has a large contrast to the gray-values surroundings. In addition, Fig. 10.5b shows the ability of the Mahalanobis classification in determining the level of oil spill spreading. In this context, this classification uses the training data by estimating means and variances of the classes, which are used to estimate the probabilities and also consider the variability of brightness values in each class [181]. These results confirm the previous study of Brekke and Solberg et al. [176]. Further, the Mahalanobis classifier provides excellent identification of oil spill pixels as with higher accuracy level of 87.8, 92.7%, using the Mahalanobis regularized Mahalanobis classifier respectively (Fig. 10.6).

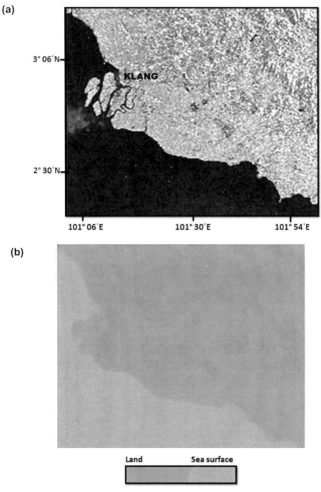

Figure 10.4. Examined RADARSAT-1 SAR (a) S2 mode and (b) Mahalanobis classifier.

10.6 Artificial Intelligent for Oil Spill Automatic Detection

The most recent work has proved that neural networks denote a competent apparatus for sculpting a diversity of non-linear discriminant complications. Therefore, neural networks, perhaps, are considered as a mathematical model comprised of numerous non-linear computational components are termed as neurons, functioning in analogous and immensely allied by associations characterized by diverse weights [172]. Artificial NNs can be depleted to acquire and replicate procedures or setups from a given oil spill in SAR images. In this context, NNs have been effectively exercised for oil spill automatic detection in SAR data. For instance, Topouzelis et al. [173], have implemented NNs to detect and classify the dark patches in SAR as an oil spill or look-alike. Generally, ANN is used to achieve two tasks: (i) dark patch detections; and (ii) physical feature detections for the oil spill and look-alikes. However, the

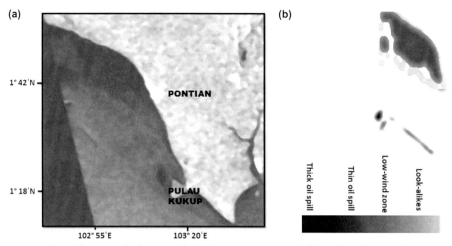

Figure 10.5. Examined RADARSAT-1 SAR (a) F1 mode and (b) Mahalanobis classifier.

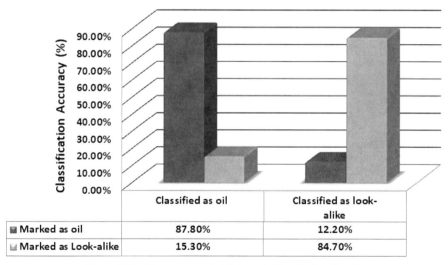

	Classified as oil	Classified as look-alike
Marked as oil	87.80%	12.20%
Marked as Look-alike	15.30%	84.70%

Figure 10.6. The accuracy of Mahalanobis Classification Results.

combination of both dark pixel detections, oil spill and lookalike discrimination by neural networks is not completely comprehended in the literature.

10.6.1 Dark Patch Detections in SAR Data

Topouzelis et al. [173] implemented a neural network backpropagation algorithm with the topology of 1:3:1. In this regard, the radar backscatter values are used as inputs to the neural networks. Moreover, the two selected training areas for each image window are selected, which are presented, the backscatter of dark patches and the surrounding sea surface. Therefore, the neural network exercises the fully connected feedforward multilayer perceptron trained by the standard backpropagation algorithm [172]. In

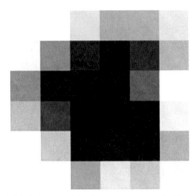

Figure 10.7. Two-bit black and white image are generated by the neural network.

these circumstances, a Log sigmoid transfer function is implemented with learning rate, momentum and the maximum number of epochs 0.05, 0.9 500, respectively. For every examined SAR data, a binary bit of black and white image which represents the dark area and the surrounding seawater is generated (Fig. 10.7).

For validation of the output of the neural networks, the reference matched image is generated using a confusion matrices (Fig. 10.8). For every generated reference image, the inclusive, dark construction and sea accuracy are estimated by the confusion matrix. In particular, accuracy is then computed by dividing the overall number of the pixels, which are properly categorized by the entire number of the pixels of the sample. In addition, dark patch precision is estimated by dividing the number of pixels perfectly categorized as a dark area by the entire number of the pixels referenced as a dark area. Sea surface pixel accuracy then is computed comparable to the dark patch accuracy, considering the overall number of pixels.

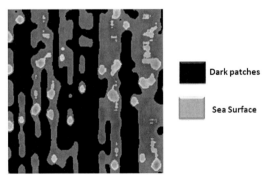

Figure 10.8. The reference image, which is generated by confusion matrices.

10.6.2 Selected Discrimination Features between Oil Spill, Look-alikes and Surrounding Sea Surface

In order to distinguish between real oil-slicks, look-alikes and varied features, some mainly used features are listed in Table 10.2. It is also very necessary to select the

Table 10.2 Mainly used characters for SAR oil-Spills detection.

Intensity	Morphology	Texture (GLCM*)
Slick Backscattering (μ_{obj})	Area (A)	Homogeneity
Slick Std. Deviation (σ_{obj})	Perimeter (P)	Contrast
Surrounding Backscattering (μ_{sce})	Complexity (C)	Dissimilarity
Intensity Ratio (μ_{obj}/μ_{sce})	Asymmetry	Entropy
Intensity St. Dev Ratio ($\sigma_{obj}/\sigma_{sce}$)	Euler number	Mean value
ISRI (μ_{obj}/σ_{obj})	Shape Index	St. Deviation
ISRO (μ_{obj}/σ_{sce})	Axis Length	Correlation
Min Slick Value (MSV)	Compactness	
Max Contrast ($\sigma_{sce}-MSV$)		
Edge Gradient		

*GLCM: Gray Level Co-occurrence Matrix

most suitable features for oil-slick classification. Though, the optimized feature sets vary along with real situations and correlate to adapt the classification algorithm, for instance, neural network [172].

Along with Table 10.2, eight parameters have coded into the genetic algorithm, which are [176]:

1. Area (A) is in km^2 of the dark SAR pixels.
2. Perimeter (P) represents the Length (in km) of the dark patches.
3. Complexity (C), which is defined as einging $C = \dfrac{P_e}{2\sqrt{\pi A}}$. In this view, complexity has a smaller value if the oil-covered has a simple geometry and larger value for irregular and complicated oil-covered.
4. Spreading (S) is derived from the principal component analysis of the vector whose components are the coordinates of the dark pixels. Consequently, two eigenvalues λ_1 and λ_2 are obtained and associated with the estimated covariance matrix under the circumstance of $\lambda_1 > \lambda_2$. The spreading rate (S) is formulated as:

$$S = 100\lambda_1 \left[\lambda_1 + \lambda_2\right]^{-1} \qquad (10.6)$$

Equation 11.1 denotes that the long and thin oil slick has a lower S value, while complicated and irregular oil slick pattern has a higher spreading value (S).

5. Oil slick Standard Deviation (OSD) is in dB, which presents the intensity values of the pixels belonging to the oil-covered.
6. Background Standard Deviation (BSD), i.e., the surrounding sea, is also in dB, which presents the intensity values of the pixels belonging to the region of interest, that is designated by the user.
7. Max Contrast (ConMax) is a difference in dB between the background mean value and the lowest value inside the oil-covered.
8. Mean Contrast (ConMe) is also a difference in dB between the background mean value and the object mean value [172].

These feature values are coded into the neural networks between 0 and 1 to avoid any bias information. In this regard, the 14-14-5-1 topology is set up for the second network, which is composed of feedforward multi-layer perceptron and trained by the backpropagation algorithm. Moreover, a Log sigmoid transfer function is used and the learning rate is set to 0.03, momentum to 0.9 and the maximum number of epochs to complete is 110. The neural network is trained with the backpropagation variant using the training set of 45 look-alikes and 35 oil spills. The test set is then classified using the set of weights previously discovered. Note that in this phase the operation is done on objects rather than pixels [173,183].

10.7 Frame Structure of Neural Network for Oil Spill Automatic Detection

The general framework of the AI Techniques used in this work is elaborated in Fig. 10.9. $\beta_1, \beta_2, \ldots, \beta_n$ are assumed as the input backscatter values from SAR data of different dark patch pixels. Figure 10.10 shows the input vectors for the Artificial Neural Network (ANN) is backscatter variation values from dark spot patches and their surrounding environment pixels in SAR data, while the output layer is oil spill or look-alikes. The output is the automatic level detection in the range between 0 and 1. In this context, oil spill pixels are represented by 1, while look-alikes or low wind zone pixels are represented by 0 [173,176].

The Neural Network (NN)-based pattern-recognition approach for Static Security Assessment (SSA) depends on the assumption that there are some characteristics of pre-contingency system states that give rise to oil spill occurrences in RADARSAT-1 SAR data which is represented by the post-contingency system. The task of the NN is to capture these common underlying characteristics of a set of known operating states and to interpolate this knowledge to classify a previously unencumbered state. The first step in such an application is to obtain a set of training data which represents the different backscatter value variations in RADARSAT-1 SAR data. The mathematical description of the above mentioned neural network is formulated as:

$$f(\mathbf{x}) = \sigma(w_{jk}\,\sigma(\Sigma_i\, w_{ij}x_i + w_{0j}) + w_{0k}) \tag{10.7}$$

where x is the input vector, w_{0j} and w_{0k} are the thresholds, w_{ij} are the weights connecting the input with the hidden nodes w_{jk} are the weights connecting the hidden with the output nodes

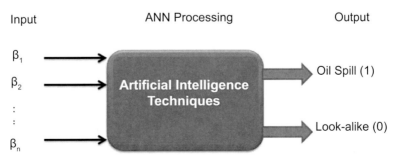

Figure 10.9. The general framework of AI Techniques.

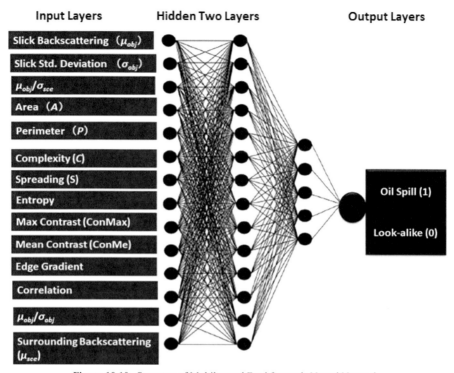

Figure 10.10. Structure of Multilayered Feed forwards Neural Network.

σ is the sigmoid activation function. These are the hidden units that enable the multilayer network to learn complex tasks by extracting progressively more meaningful information from the input examples. The multilayer network MLP has a *highly connected topology* since every input is connected to all nodes in the first hidden layer, every unit in the hidden layers is connected to all nodes in the next layer, and so on. The input features, initially these are the input examples, propagate through the neural network in a forward direction on a layer-by-layer basis, that is why they are often called feedforward multilayer networks [173] [186].

10.8 Backpropagation Learning Algorithm for Automatic Detection of Oil Spill

MLP is successfully applied to the SAR image after acquiring a supervised training algorithm for learning their weights through the *backpropagation learning algorithm*. The backpropagation algorithm for training multilayer neural networks is a generalization of the LMS training procedure for nonlinear logistic outputs. As with the LMS procedure, training is iterative with the weights adjusted after the presentation of each case. The error backpropagation algorithm contains dual passes over the network forward pass and backward pass. During the backward pass, the weights are adjusted in accordance with the *error correction rule*. It suggests that the actual network output is subtracted from the given output in the SAR image. The weights are adjusted so as to make the network output closer to the desired one.

The backpropagation algorithm does gradient descent as it moves in a direction opposite to the gradient of the error, that is in the direction of the steepest decrease of the error. This is the direction of most rapid error decrease by varying all the weights simultaneously in the proportion of how much good is done by the individual changes [186]:

$$\nabla E(\mathbf{w}) = [\partial E/\partial w_0, \partial E/\partial w_1, \ldots, \partial E/\partial w_d] \tag{10.8}$$

Equation 10.8 reveals that the gradient descent learning strategy requires a continuous activation function. The network is trained initially selecting small random weights and then presenting all training data incrementally. Weights are accustomed after every trial using side information specifying the correct class until weights convergence and the cost function is reduced to an acceptable value. In this case, the training algorithm for multilayer networks necessitates differentiable, continuous nonlinear activation functions. Such a function is the sigmoid or logistic function:

$$O = \sigma(s) = 1/(1 + e^{-s}), \tag{10.9}$$

where s is the sum: $s = \Sigma_{i=0}^{d} w_i x_i$. Sometimes σ is known as an alternatively squashing function, for instance, it records a very large input domain to a small range of outputs. On the contrary, a nonlinear function frequently exploits in practice the *hyperbolic tangent*:

$$O = \tanh(s) = (e^s - e^{-s})/(e^s + e^{-s}) \tag{10.10}$$

Every so often, the hyperbolic tangent is chosen as it renders the training a little easier. Therefore, the major steps in the training algorithm are feedforward calculations, propagating error from the output layer to the input layer and weight updating in hidden and output layers [186]. Forward pass phase calculations are shown by the following equations between input (i) and hidden (j) [183].

$$\theta_j = \frac{1}{1 + e^{(\Sigma_j w_{ij}\theta_i + \theta_j)}} . \tag{10.11}$$

$$\theta_k = \frac{1}{1 + e^{(\Sigma_k w_{jk}\theta_j + \theta_k)}} E. \tag{10.12}$$

where θ_j is the output of node j, θ_i is the output of node i, θ_k is the output of the node w_{jk} is the weight connected between node i and j, and θ_j is the bias of node j, θ_k is the bias of node k. In the backward pass phase, error propagated backward through the network from the output layer to the input layer as represented in Equation 10.12. Following Topouzelis et al. [22]. The weights are modified to minimize the Mean Square Error (MSE).

$$MSE = \frac{1}{n}\sum_{i=1}^{n}\sum_{j=a}^{m}(d_{ij} - y_{ij})^2. \tag{10.13}$$

where d_{ij} is the jth desired output for the ith training pattern, and y_{ij} i. The corresponding actual output [182,186]. Finally, the Receiver-Operator-Characteristics (ROC) curve and error standard deviation are used to determine the accuracy level of each algorithm that has been used in this study. In addition, ROC and error standard deviation was used to determine the accuracy of feature detections in RADARSAT-1 SAR data.

10.9 Backpropagation Training Algorithm

Initialization: Examples $\{(\mathbf{x}_e, y_e)\}_{e=1}^N$, initial weights w_i set to small random values, learning rate $\eta = 0.1$

Repeat

  for each training example $(\mathbf{x}, y)$

  - *calculate the outputs* using the sigmoid function:

  $o_j = \sigma(s_j) = 1/(1 + e^{-sj})$, $s_j = \Sigma_{i=0}^d w_{ij} o_i$ where $o_i = x_i$ // *at the hidden units j*
  $o_k = \sigma(s_k) = 1/(1 + e^{-sk})$, $s_k = \Sigma_{i=0}^d w_{jk} o_j$ // *at the output units k*

  compute the benefit βk **at the nodes** k **in the output layer:**

  $\beta_k = o_k (1-o_k) [y_k - o_k]$ // *effects from the output nodes*

  compute the changes for weights $j{\rightarrow}k$ **on connections to nodes in the output layer:**

  $\Delta w_{jk} = \eta \ \beta_k o_j$ // *effects from the output of the neuron*
  $\Delta w_{0k} = \eta \ \beta_k$

  compute the benefit β_j **for the hidden nodes** j **with the formula**

  $\beta_j = o_j (1-o_j) [\Sigma_k \beta_k w_{jk}]$ // *effects from multiple nodes in the next layer*

  compute the changes for the weights $i{\rightarrow}j$ **on connections to nodes in the hidden layer:**

  $\Delta w_{ij} = \eta \ \beta_j o_i$ // *effects from the output of the neuron*
  $\Delta w_{0j} = \eta \beta_j$

  - *update the weights* by the computed changes:

  $w = w + \Delta w$

  until termination condition is satisfied.

10.10 Oil Spill Detection by Neural Network Algorithm

It is worth noting that the neural network algorithm is able to isolate oil spill dark pixels from the surrounding environment. In other words, look-alikes, low wind zone, sea surface roughness and land are marked by the white color while oil spill pixels are marked all black. Figure 10.11 reveals an excellent pattern recognition of the oil spill from W1 mode data. Moreover, the dark patches have broken into several pieces as a result of the impact of Langmuir circulations, which is generated by the formation of cells of rotating water that parallel the surface under wind and current effects.

Like W1 mode, F1 mode data also reveal the perfect morphological pattern of the oil spill (Fig. 10.13). On top of that, the neural network algorithm can decide the existence of the oil spill in the SAR image. In this view, the neural network does not assign any particular dark patches in S2 mode data due to the absence of the oil spill event. In this case, the neural network can distinguish between land and sea surface as shown in Fig. 10.12.

It is interesting to find that 99% of the oil spill in the test set were correctly classified. Three scenes by the leave-one-out method presented an exact classification of 99% for oil spills (an approach based on multilayer perceptron (MLP) neural network with two hidden layers). The net is trained using the back-propagation algorithm to

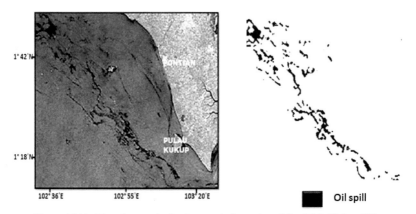

Figure 10.11. Neural networks for Automatic Detection of the Oil Spill from W1.

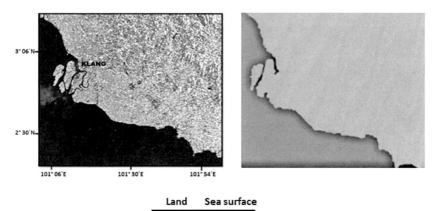

Figure 10.12. Neural networks for automatic detection of land and sea surface from S2 data.

Color version at the end of the book

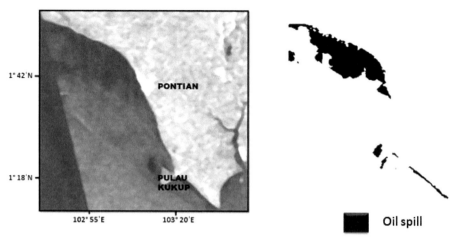

Figure 10.13. Neural networks for automatic detection of the oil spill from F1 mode data.

minimize the error function. Ninety nine percent of oil spills are automatically detected using the leave-one-out method. This study agrees with the study of Topouzelis et al. [173].

10.11 Comparison between Mahalanobis Classifier and Neural Networks

The Receiver–Operator Characteristics (ROC) curve in Fig. 10.14 indicates a significant difference in the discrimination between oil spill, look-alikes and sea surface roughness pixels. In terms of ROC area, the oil spill has an area difference of 20 and 35% for a look-alike and 30% for sea roughness and a ρ value below 0.005 which confirms the study of Marghany et al. [184,185]. This suggests that the Mahalanobis classifier and neural networks are good methods to discriminate region of oil slicks from surrounding water features.

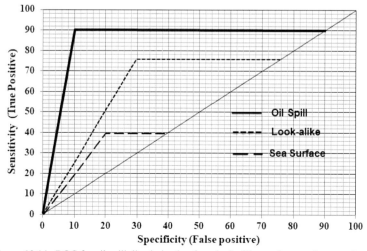

Figure 10.14. ROC for oil spill discrimination from look-alikes and sea surface roughness.

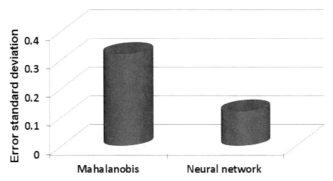

Figure 10.15. Standard Error for different algorithms.

Figure 10.15, however, shows the standard deviation of the estimated error for neural network of value 0.12 is lower than the Mahalanobis one. This suggests that ANN performed accurately as an automatic detection tool for the oil spill in RADARSAT data. The good performance of the neural algorithm encouraged a second phase where optimization of the net from the point of view of the number of its adaptive parameters (units and connections) has been carried out by using a pruning procedure. Accordingly, a network is examined to assess the relative importance of its weights, and the least important ones are deleted. Typically, this is followed by some further training of the pruned network, and the procedure of pruning and training may be repeated for several cycles. Clearly, there are various choices to be made concerning how much training is applied at each stage, what fraction of the weights is pruned, and so on. In the present work, every time the weight was removed, the new net is generated is trained, as in the case of the initial training. The overall error value approaches a value of convergence, and, since the study started with a net committing no errors, the pruning procedure was continued until it was realized that new removals involve errors in the classification task. The most important consideration, however, is how to decide which weights should be removed [21]. To do this, some measure of the relative importance was needed, or saliency of different weights. This result agrees with Topouzelis et al. [20,22].

Figure 10.16 shows that W1 mode data has a lower percentage value of the standard error of 15% in comparison to F1 and S2 mode data. This means that W1 mode performs better detection of oil spills than F1 and S2 modes. In fact, the W1 mode showed the steeper incident angle of 30° than the S1 and S2 modes. The offshore wind speed during the W1 mode overpass was 4.11 m/s, whereas the offshore wind speed was 7 m/s during the S2 mode overpass. Wind speeds below 6 m/s are appropriate for detection of oil spill in SAR data [176]. Therefore, applications requiring imaging of ocean surface, steep incidence angles are preferable as there is a greater contrast of backscatter manifested at the ocean surface [185].

In contrast to previous studies of Fiscella et al. [177] and Marghany et al. [185], the Mahalanobis classifier provides a classification pattern of an oil spill where the slight oil spill can distinguish from medium and heavy oil spill pixels. Nevertheless, this study is consistent with Topouzelis et al. [173]. In consequence, the ANN extracted oil spill pixels automatically from surrounding pixels without using a thresholding

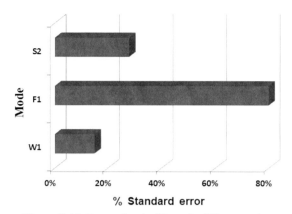

Figure 10.16. Percent Standard Error for different modes.

technique or different segmentation algorithm as stated at Solberg et al. [176] and Samad and Mansor [180].

Finally, the results show that ANN distinguishes an oil spill from the surrounding sea surface features with a standard deviation of 0.12. In conclusion, integration of different algorithms for oil spill detection provides better ways of getting effective oil detection. The ANN algorithms is an appropriate algorithm for oil spill detection and while the W1 mode is appropriate for the oil spill and look-alikes discrimination and detection using RADARSAT-1 SAR data.

Fractal Algorithm for Discrimination Between Oil Spill and Look-Alike

11.1 Definitions of Fractal

The fractal word is derived from the Latin word fractus signifying "irregular segments". Mandelbrot was the pioneer scientist, who defined and explored the fractal in 1977. In mathematics, a fractal is a subset of a Euclidean dimension for which the Hausdorff dimension (Fig. 11.1) rigorously surpasses the topological space. Fractals develop to exhibit almost the identical at exclusive levels, as is illustrated here in the successively small magnifications of the Mandelbrot set (Fig. 11.2) [187–190]. Because of this, fractals are encountered ubiquitously in nature. Fractals exhibit similar patterns of an increasing number of small scales referred to as self-similarity [188], additionally recognized as increasing symmetry or unfolding symmetry; If this replication is precisely identical at each and every scale, as in the Menger sponge [190], it is known as affine self-similar. Nonetheless, scientists disagree on the exact definition of the fractal. In this sense, the majority agrees with the rudimentary concepts of self-similarity and the infrequent correlation fractals have with space they are implanted in [187],189,191,192].

Fractal, therefore, is the standard mathematical method that is required for a high level of image processing. Indeed, fractal geometry can be used to discriminate between different textures. Consistent with Redondo [193], fractal refers to entities, especially sets of pixels, that display a degree of self-similarity at different scales. Consequently, self-similarity is the foundation for fractal analysis, which is applied to a group of pixel intensities with the same trend of variation [194]. It is defined as a property of a curve or surface where each part is indistinguishable from the whole, or where the form of the curve or surface is invariant with respect to scales. In this case, the curve or surface is made of copies of itself at different scales [196].

Figure 11.1. The circle fractal with Hausdorff dimension.

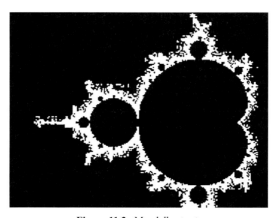

Figure 11.2. Mandelbrot set.

11.2 Fractal Dimensions

One point agreed on is that fractal patterns are characterized by fractal dimensions, but whereas these numbers quantify complexity (i.e., changing detail with changing scale), they are neither unique nor specify details that describe how to construct particular fractal patterns [193].

If the real-world images or object retain the fractal characteristics, it can be assumed as an existence created of an inestimable quantity of reproductions of itself, fluctuating in size from huge to tiny. Nonetheless, real-world images (or objects) are not accurately fractal; hereafter, the multifractal perception was familiarized. The image or object is deliberated to be self-possessed of a finite quantity of subclasses, and its multifractal dimension contemplates individually a subset's characteristic and is consequently signified to be an extra dominant quantity of image texture. In this regard,

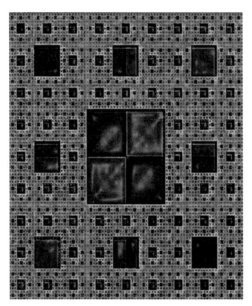

Figure 11.3. Menger sponge fractal.

the fractal dimension of real-world images or object is a quantity of its complexity. In this view, the higher the fractal dimension, the more complex is real-world images or the object nature or assembly (Fig. 11.4). In other words, the fractal dimension can consequently be utilized as a magnitude of texture quantization, identical to complex surfaces. In this understanding, the image texture is a function of intensity variations. Therefore the rough texture has a high intensity, which indicates a rough surface, while the smooth surface with low intensity presents a fine texture [188–191].

Consistent with Falconer [196], fractals should, in addition to being nowhere differentiable and able to have a fractal dimension, be only generally characterized by a gestalt of the succeeding features self-similarity, fine or detailed structure at arbitrarily small scales, Irregularity locally and globally, and simple and "perhaps recursive". In this regard, self-similarity involves qualitative self-similarity, for instance, as in a time series. On the contrary, statistical self-similarity duplicates a pattern stochastically (Fig. 11.4). In this view, numerical or statistical measures are preserved across scales, for instance, the coastline of Britain whose length has to be measured on a map using a measuring instrument that has a specific step length. Moreover, exact self-similarity: identical at all scales, for instance, Koch snowflake (Fig. 11.5), in which its segment is scaled and repeated as neatly as the repeated unit that defines fractals. Conversely, quasi self-similarity, which approximates the similar shape at diverse scales; may enclose small duplicates of the entire fractal in distorted and degenerate forms; for instance, the Mandelbrot set's satellites. In this sense, they are approximations of the entire set, but not exact copies. Finally, multifractal scaling is one feature of self-similarity which is characterized by more than one fractal dimension or scaling rule.

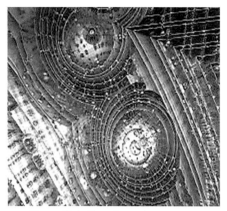

Figure 11.4. Fractal complex structure.

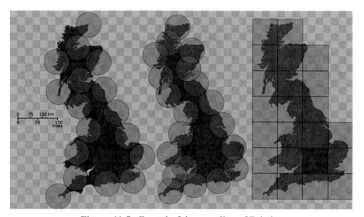

Figure 11.5. Fractal of the coastline of Britain.

11.3 Estimation of Fractal

Let assume that T, D, and H are the topological dimension, fractal dimension and the Hurst exponent, respectively. The mathematical estimations of H and D can be formulated as:

$$E[\Delta^2 f] = a[\Delta^H d]^2 \tag{11.1}$$

where $E, \Delta f, \Delta d$, and a are the expectation operator, intensity operation, spatial distance and scaling constant. Substitute, therefore, $H=3\text{-}D$, and $\kappa = E(|\Delta f|)_{\Delta d = 1}$ in the Equation (11.1), and we obtain the following formula:

$$E(|\Delta f|) = \kappa \Delta d^H \tag{11.2}$$

Let us apply the log to both sides of Equation 11.2, then

$$\log E(|\Delta f|) = \log \kappa + H \log(\Delta d) \tag{11.3}$$

Figure 11.6. Koch snowflake fractal pattern.

The Hurst exponent H can be obtained by using the least-squares linear regression to estimate the slope of the gray-level difference $GD(k)$ versus k in log-log scales and k varies from 1 to the maximum value where

$$GD(k) = \frac{\sum_{i=1}^{N}\sum_{j=1}^{N-k-1}\left|I(i,j) - I(i, j+k)\right| + \sum_{i=1}^{N-k-1}\sum_{j=1}^{N}\left|I(i,j) - I(i+k, j)\right|}{2N(N-k-1)} \qquad (11.4)$$

The fractal dimension D can be derived from the relation $D=3-H$. The approximation error of the regression line fit should be determined to prove that the analyzed texture is fractal, and thus be efficiently described using fractal measures. A small value of the fractal dimension D implies to a large value of the Hurst exponent H represents a fine texture, while a large D, implies to a smaller H value, corresponds to the coarse texture.

11.4 Estimation of Hurst Exponent

To calculate approximately the Hurst exponent, initially, one must first estimate the dependence of the rescaled range R must be estimated as a function of n observation. Let L which is the time series of full-length L, that is divided into a number of shorter time series of length $n = L, L/2, L/4, ...$ The average rescaled range is then calculated for each value of n.therefore, R for a (partial) time series of length n, $X_1, X_2, ..., X_n$ is calculated as follows:

1. Estimate the mean (μ)

$$\mu = n^{-1}\sum_{i=1}^{n} X_i \qquad (11.5)$$

2. Generate a μ_t-regulated series

$$\mu_t = X_t - \mu \quad \text{for } t = 1, 2, ..., n \tag{11.6}$$

Then the cumulative deviate series Z_t can be computed by:

$$Z_t = \sum_{i=1}^{t} \mu_i \quad \text{for } t = 1, 2, ..., n \tag{11.7}$$

Then the R can be estimated as follows:

$$R(n) = \max(Z_1, Z_2, ..., Z_n) - \min(Z_1, Z_2, ..., Z_n) \tag{11.8}$$

Therefore, the standard deviation σ of $X_1, X_2, ..., X_n$ can be given by:

$$\sigma = \sqrt{n^{-1} \sum_{i-1}^{n} (X_i - \mu)^2} \tag{11.9}$$

The Hurst exponent is estimated by fitting the power law $E\left[\dfrac{R(n)}{\sigma(n)}\right] = an^H$ to

the SAR data. Therefore, this can be achieved by plotting $\log\left[\dfrac{R(n)}{\sigma(n)}\right]$ as a function

of $\log[n]$, and fitting a straight line; the slope of the line obtains H. Such a diagram is termed a box plot. Nonetheless, this technique is well-known to create biased approximations of the power-law exponent. For small n, there is a significant deviation from the 0.5 slopes. The theoretical mathematical correction of these biases can be formulated as:

$$E\left[\frac{R(n)}{\sigma(n)}\right] = \begin{cases} \left[\left(n(0.5\pi)\right)^{-0.5}\right] \sum_{i=1}^{n-1} \sqrt{\dfrac{n-i}{i}}, & \text{for } n > 340 \\[4mm] \dfrac{\Gamma\left(\dfrac{n-1}{2}\right)}{\sqrt{\pi}\,\Gamma(0.5n)} \sum_{i=1}^{n-1} \sqrt{\dfrac{n-i}{i}}, & \text{for } n \le 340 \end{cases} \tag{11.10}$$

where Γ is the Euler gamma function which is given by:

$$\Gamma(n) = (n-1)! \tag{11.10.1}$$

The correction of the Hurst exponent $\dfrac{R(n)}{\sigma(n)}$ is calculated as 0.5 plus the slope

of $\dfrac{R(n)}{\sigma(n)} - E\left[\dfrac{R(n)}{\sigma(n)}\right]$ [203].

11.5 Fractal Algorithm for Oil Spill Identification

The oil slick detection tool uses fractal algorithms to detect self-similarity characteristics. A box-counting algorithm introduced by Benelli and Garzelli [197] divided an irregular shape of the slick in the RADARSAT-1 SAR image plane (i,j), into smaller boxes (Fig. 11.7). This was done by dividing the initial length of the convoluted slick line at the complexity level C_s by the recurrence level of the iteration [198]. Complexity (C_s), therefore, which is defined as:

$$C_s = \frac{P_e}{2\sqrt{\pi A}}. \tag{11.11}$$

In this view, complexity has a smaller value if the oil-covered has a simple geometry and larger value for irregular and complicated oil-covered. Area (A) is in km² of the dark SAR pixels and Perimeter P_e represents the Length (in km) of the dark patches. In this regard, we define a decreasing sequence of Complexity C_s tending from ComplexityC_0, the largest value, to less than or equal to zero. The fractal dimension $D\,(C_s)$ as a function of the SAR image complexity C_s is given by:

$$D(C_s) = \lim_{S \to \infty} \frac{logM(C_s)}{-log(C_s)} \tag{11.12}$$

where $M(C_s)$ denotes the number of boxes needed to cover the various slick areas with different complexity C_s in the SAR images. The number of boxes of side length l_s needed to cover a fractal profile, varies as C_s^{-D}, where D is the fractal dimension that is to be estimated. If the sampled profile is a fractal object, then $M(C_s)$ is proportional to C_s^{-D}. Therefore, the following relation, adopted from Milan et al. [199], should be satisfied:

$$M(C_s) = aC_s^{-D} \tag{11.13}$$

where a is a positive constant, derived from a linear regression analysis between log $M(C_s)$ and log (C_s). For different box sizes C_s, points are plotted in the log-log plane. The dimension $D\,(C_s) = D_B$ can then be estimated [199].

Consistent with Sarkar and Chaudhuri [200], a complication occurs when computing $D(C_s)$ with Equation 11.12, due to the discrete SAR image surfaces. Therefore, approximations to this relationship are used. First, the SAR intensity image

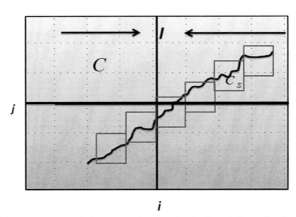

Figure 11.7. Box-counting technique for estimating the the fractal dimension from SAR data.

is treated as a two-dimensional matrix ($C \times C$). This $C \times C$ complexity image matrix is then divided into overlapping or abutted windows of size $l_s \times l_s$. For each window, there is a column of accumulated boxes, each of the size of $l_s^2 \times l$. The complexity values C_0 are stored at each intersection of column i and row j of the various slick areas (Fig. 11.7). Then l is calculated by using the differential box-counting proposed by Sarkar and Chaudhuri [200]:

$$\left[\frac{C_s}{l} \right] = \left[\frac{C_0}{l_s} \right] \tag{11.14}$$

Let the minimum and maximum (C_s) in the (i, j) window fall into boxes numbered n and m. The total number of boxes needed to cover the various slick pixels in the SAR image with the box size $l_s^2 \times l$ equals:

$$M(C_s) = \sum_{i,j}^{l} n(C_0) - m(C_s) + 1 \tag{11.15}$$

Let $P[M(C_s), l_s]$ be the probability of the total number of the box $M(C_s)$ with box sizes l_s. This probability should be directly proportional to the number of boxes $\sum_{i,j}^{l} n(C_0) - m(C_s) + 1$ spanned on the (i, j) windows. By using Equation (11.15) the expected number of boxes with the size l_s needed to cover the slick pixels can be calculated using:

$$M(C_s) = \sum_{i,j} \frac{1}{n} P[M(\beta_s), l_s] \tag{11.16}$$

Consistent with Fiscella et al. [177], the probability distribution of the dark area belonging to slick pixels can be calculated using the formula:

$$P[M(C_s)] = [1 + \prod_n q_n(M(C_s)) / p_n(M(C_s))] \tag{11.17}$$

Let $n = \sum_{i,j}^{l} n(C_0) - m(C_s) + 1$, and let q and p be the probability distribution functions for look-alike and oil spill pixel areas, respectively. From Equations (11.16), (11.17) and (11.12) one obtains a new formula for estimating the fractal dimension D_B

$$D(C_s) = D_B = \lim_{s \to \infty} \frac{\log \sum_{i,j} n^{-1}[1 + \prod_n q_n(M(C_s)) / p_n(M(C_s))]}{-\log(C_s)} \tag{11.18}$$

In practice, the limit for $M(C_s)$ going to zero cannot be taken as it does not produce a texture image for oil spills or look-alikes in SAR data. Using fractal dimensions to quantify texture for segmentation, we divide the whole SAR data into overlapping sub-images. Each sub-image is centered on the pixel of interest. We then estimate the fractal dimension $D(C_s)$ within each sub-image and assign the fractal dimension value to the central pixel of each sub-image. The threshold is then applied in which it relies on the resulting fractal map values. This produces a texture image that may be used as an additional feature in a slick pixel classification.

11.6 Otsu Thresholding Algorithm

The automatic detection of the oil spill can be achieved by image segmentation. In fact, image segmentation divides the complex fractal image (11.8) into several segments (Fig. 11.8). In other words, segmentation is implemented to simplify and alternate the illustration of satellite data to acquire an accurate image analysis . In this regard, unsupervised Otsu's thresholding technique is taken into consideration to partition an image into non-intersecting clustering. Otsu's approach, therefore, is automatically detected in several clusters, which, is based on image thresholding or reduction of the complexity level image to a binary image. Formerly, histogram probability thresholding locates the specific object from the heritage.

In Otsu's approach, we systematically explore the threshold, which, minimizes the variance within the cluster, expressed as a weighted sum of variances of the two clusters, which, are defined as:

$$\sigma_w^2(t) = D(C_s)_1(t)\sigma_1^2(t) + D(C_s)_2(t)\sigma_2^2(t) \tag{11.19}$$

here σ_1 and σ_2 are variances of the two clusters which are derived by $D(C_s)$, and t is a threshold, which is estimated from L bins of the histogram. In other words, Otsu's approach can be formulated in terms of cluster means μ as:

$$\mu_1(t) = \frac{\sum_{i=0}^{t-1} ip(i)}{\sum_{i=1}^{t} p(i)} \tag{11.20}$$

$$\mu_1(t) = \frac{\sum_{i=0}^{L} ip(i)}{\sum_{i=t+1}^{L-1} p(i)} \tag{11.21}$$

being μ_1 and μ_2 are the means of $D(C_s)$ based on the probability, respectively. Equations 11.20 and 11.21 demonstrate that the cluster probabilities and their means can be

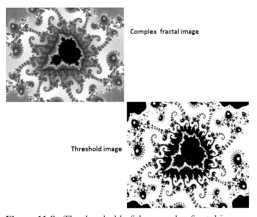

Complex fractal image

Threshold image

Figure 11.8. The threshold of the complex fractal image.

```
Begin
Do for L = 2 in M(C_s)/2
Define box size l_1^2 × l
Partition the image into overlapping windows of size l_1^2 × l
Do for each window in the image
Construct a column of boxex and assign numbers 1, 2,…, to the
Boxes from bottom to the top;
Find the bottem to the top;
Find the box numbers wich interest the maximal and the
Minimal (Cs) values within the window;
EndDo
Calculate M(C_1) using (11.13);
EndDo
Estimating the fractal dimension D(C_s) using 11.14;
Calculated histogram and probabilities of every D(C_s) level,
Create initial D(C_s) (0) and μ_i(0)
Determine through all conceivable threholds t=1,……maximum
intensity
Appraise D(C_s) and μ_i
Calculate σ_w^2(t)
Preferred threshold matches to the maximum σ_w^2(t)
End
```

Figure 11.9. The pseudo code of the fractal dimension based on Otsu algorithm.

calculated iteratively [201]. The pseudo code of the Otsu algorithm is shown in Fig. 11.9.

Since the entire kernel windows is continuous and self-determining of t, the result of changing the threshold is merely to transfer the influences of the two terms back and forth. Hence, minimizing the within-cluster variance is similar to maximizing the between-cluster variance.

In other words, thresholding is a simple form of segmentation wherein each pixel in an image is compared with the threshold value. If the pixel lies above the threshold, it will likely be marked as foreground and if it is beneath a threshold as heritage. The threshold will most customarily be $D(C_s)$ value. In this approach, the selection of initial threshold value relies on the histogram of $D(C_s)$ of an image (Fig. 11.8). The connected element evaluation is used here to extract the nearby object shape descriptors for figuring out the preferred target. The template is implemented as an identical model. Subsequently, correlation measurement is exercised for measuring the similarity between item vicinity functions and simulation verified that the functionality of object monitoring in SAR images with the assistance of used techniques [202].

11.7 Examined SAR Satellite Data

The fractal formula is trained on five different RADARSAT-1 SAR mode data, whereas the dark spots are identified and examined. The RADARSAT-1 SAR images contain the confirmed oil spills which occurred near the west and east coasts of Peninsular

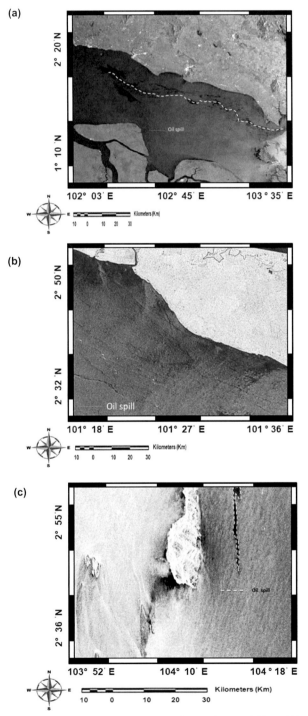

Figure 11.10. Oil spill locations are indicated by dash lines during acquisition of (a) RADARSAT-1 Wide mode (W1), (b) RADARSAT-1 Standard mode (S2) and (c) RADARSAT-1 SAR Standard mode (S1).

Malaysia on 26 December 1997 [179], 20 December 1999 [180], and 26 March 2008, respectively (Fig. 11.10).

Further, *in situ* observations have confirmed that the occurrence of oil spills on 26 March 2008 in which an area located in between 103° 52' 51.75"E to 104° 19' 13.52"E and 2° 36' 29.20"N to 2° 54' 07.33"N were covered. Therefore, the validation of new fractal formula has been examined in pair of RADARSAT-1 SAR Standard beam mode (S2) which were acquired on 15 November 1996 and 9 July 2005, respectively, in the Malacca Straits coastal waters (Fig. 11.11). Therefore, both Mohamed et al. [179] and Samad and Shattri [180] reported the occurrence of oil spill pollution on 15 December 1997 and 20 December 1999, respectively along the coastal water of the Malacca Straits. Further, the occurrence of oil spills pollution on

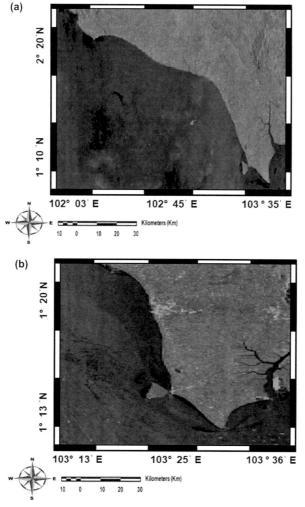

Figure 11.11. RADARSAT-1 SAR Standard beam mode (S2) which are acquired on (a) 15 November 1996 and (b) 9 July 2005.

26 March 2008 is confirmed by field work observation between 103° 52' 51.75"E to 104° 19' 13.52"E and 2° 36' 29.20"N to 2° 54' 07.33"N.

SAR data acquired in this study were derived from the RADARSAT-1 images that involve Wide beam mode (W1), Standard beam mode (S1) and Standard beam mode (S2) images, respectively. These data are C-band and have a lower signal-to-noise ratio due to their HH polarization with a wavelength of 5.6 cm and a frequency of 5.3 GHz. RADARSAT-1 wide beam mode (W1) data have four independent looks and cover incidence angles of 20° to 31° [189] whereas Standard beam mode (S1) and (S2) data have 3.1 looks and cover incidence angles of 23.7°and 31.0° [180,189]. Indeed, spatial averaging of (4-look) SAR data reduces signal level fluctuation due to speckle to about 5% which is equivalent to 0.2 dB. Therefore, a calibration accuracy of 0.2 dB appears an appropriate task for slick detection in RADARSAT-1 data [189]. In addition, Wide beam mode (W1) data and Standard beam mode (S2) cover the swath width of 165 km and 100 km, respectively. Furthermore, both data have different of an azimuth resolution and a ground range resolution (Table 11.1).

Table 11.1 RADARSAT-1 SAR Wide (W1), Standard (S2) and Standard (S1)characteristics.

Beam mode	Incidence Angle (°)	Swath area (km)	Looks	Width (km)	Resolution (Range x Azimuth, m)
W1	20–31	150	4	165	30–48 x 28
S2	23.7–31	100	3.1	100	25 x 28
S1	23.7–31	100	3.1	100	25 x 28

11.8 Backscatter, Incident Angle and Wind Variation along Suspected Oil Spill Patches

Figure 11.12 shows the variation of the average backscatter intensity along the azimuth direction in the oil-covered areas as a function of incidence angles for the W1, S2 and S1 modes, respectively. The backscattered intensity is dampened by –8 dB to –18 dB in W, –10 dB to –18 dB in S2 and –11dB to –18 dB in S1 mode data (Fig. 11.12a). However, W1, S2, and S1 mode data have a lower backscatter intensities as compared to pairs of S2 mode data (Fig. 11.12b). Further, five different mode backscatter intensities are well above the RADARSAT-1 noise floor value of nominally –20 dB. In fact RADARSAT-1 SAR is a C-band instrument with a variable acquisition swath, presenting a large variety of possible incidence angles, swath widths, and resolutions [198]. Oil slicks can be detected with a contrast as small as 4 dB [198,204,180]. This suggests that a large part of the RADARSAT-1 swath could be useful for oil slick detection. Nevertheless, Ivanov et al. [205] reported that the RADARSAT-1 SAR, in its ScanSAR Narrow mode with a swath width above 300 km, is attractive for marine oil pollution detection.

The wind speed conditions acquired from the Malaysian Meteorological Survey Department showed a maximum offshore wind speed velocity of 4.11 m/s during the W1 data overpass and of 5.4 and 6.8 m/s during the acquisition of S2 and S1 mode data, respectively (Fig. 11.12a). Further, maximum offshore wind speed velocity of

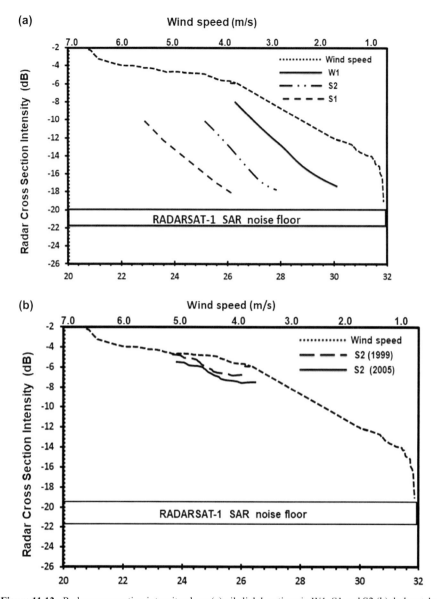

Figure 11.12. Radar cross section intensity along (a) oil slick locations in W1, S1 and S2 (b) dark patches in pair of S2 beam mode data (1999 and 2005) and wind speed distribution during the date of acquisitions.

5 m/s was noticed during the pair of S2 mode data overpasses on 15 November 1996 and 9 July 2005, respectively, in the Malacca Straits coastal waters (Fig. 11.12b). In addition, the oil spill in the W1 mode with steeper incidence angle is between 26° and 30° (Fig. 11.12a), whereas in the S2 and S1 mode data the oil spills are imaged by shallower incidence angle between 23° and 27°. Consistent with Marghany et al. [198] for oil spill detection, steeper incidence angles are preferred since they tend to

maximize the signal from the ocean surface. These results of backscatter variations across oil spill locations agree with the study of Marghany et al. [206].

11.9 Fractal Map of Oil Spill and Look-alikes

The proposed method to estimate the fractal dimension has been applied to the amplitude RADARSAT-1 SAR data by using a 10 x 10 block in the RADARSAT-1 SAR data (Fig. 11.13). Figure 11.13 shows the selected zoom window in RADARSAT-1 SAR S2 mode data of Fig. 11.10b. It is clearly noticed that the box-counting size of 10 x 10 is appropriate for feature discriminations in RADARSAT-1 data as compared to box counting sizes of 9 x 9 and 11 x 11, respectively. In fact, the different information mapped by the fractal dimension of a box counting size 10 x 10 are preserved and clearly identified. This result confirms the study of Gade and Redondo (1999). The fractal dimension maps show good discrimination between different textures on the RADARSAT-1 SAR images and correlate well with image texture regions. This is clearly noticed at the area (H) where the ship and wake are well identified (Fig. 11.13).

The oil spill pixels are dominated by lower fractal values than look-alikes and the surrounding environments (Fig. 11.14). In Fig. 11.14a, the fractal values of oil spill regions vary between 1.48 and 2. According to Marghany et al. [185], the oil spill becomes thinner when the fractal dimension value increases. This can be noticed in areas A to C. In fact, a thick oil spill dampens small scale waves and therefore there is no Bragg resonance, which reduces the roughness of sea surface as compared to a thin oil spill [206, 208]. In this context, the fractal dimension is a function of

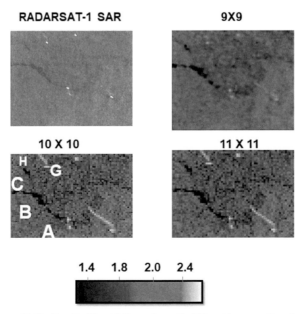

Figure 11.13. Result of fractal dimension with different box counting sizes.

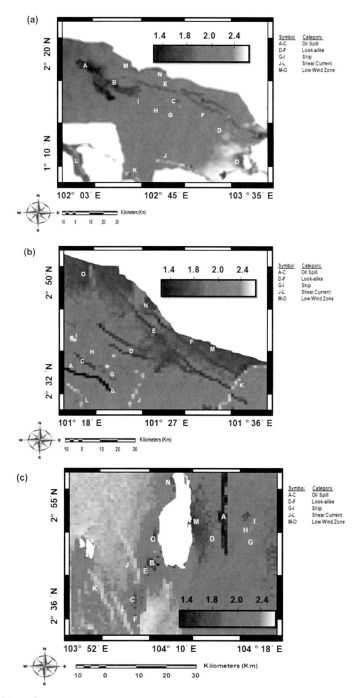

Figure 11.14 contd. ...

... Figure 11.14 contd.

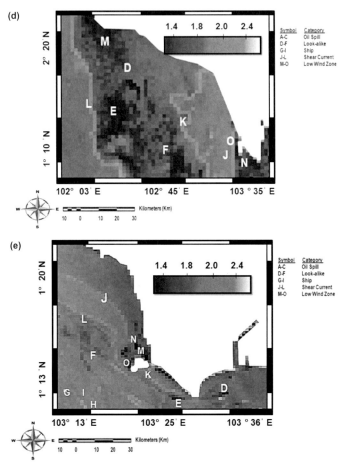

Figure 14.14. Fractal map for RADARSAT-1 SAR (a) Wide mode (W1), Standard mode (S2), Standard mode (S1) and pair of Standard mode (S2) acquired on (d) 1999 and (e) 2005.

sea surface level intensities over the RADARSAT-1 SAR images which express the self-similarity (Benelli and Garzelli (197). In Figs. 11.14b and 11.14c, however, a reduction in the maximum fractal dimension at area C in S1 and S2 data respectively. This is probably caused by the short span of oil spill in the S1 and S2 mode data and could be attributed to the spatial variation of S1 and S2 mode data reflectivity as a function of shallower incidence angle (Fig. 11.12) and smaller swath area as compared to the W2 mode data (Table 11.1).

In contrast to the W1 mode data, the fractal dimension values of look-alikes in both S1 and S2 mode data are higher. In the S1 and S2 mode data, areas F and E have represented the occurrence of look-alikes. Table 11.2 shows that area F in the S1 and S2 mode data corresponds to fractal dimension value 1.93 and 2.0, respectively, whereas area E corresponds to fractal dimension equal to 1.72 in S1 mode data and

Table 11.2. Fractal values for different features in RADARSAT-1 SAR Wide (W1) and Standard (S2).

Area	Fractal dimension		
	W1	**S2**	**S1**
Oil spill			
A	1.48	1.49	1.48
B	1.52	1.52	1.50
C	1.40	1.57	1.60
Look-alike			
D	1.5	1.52	1.50
E	1.6	1.70	1.72
F	1.8	1.8	1.82
Ship			
G	2.9	2.9	3.0
H	2.6	2.4	2.3
I	2.9	3.0	2.8
Shear current			
J	2.7	2.8	2.8
K	2.8	2.9	2.9
L	2.9	2.9	2.9
Low wind zone			
M	1.71	1.57	1.75
N	1.72	1.60	1.72
O	1.74	1.64	1.33

1.7 in S2 mode data. In the three modes, the highest fractal dimension values of 2.8 and 3.0 in areas I and G, respectively, are represented by the presence of a ship, whereas ship waves have a lower fractal dimension values between 1.3, 1.4 and 1.6 in area H in S1,S2 and W1 mode data, respectively (Table 11.2).

Furthermore, the occurrence of shear current flow can be seen in areas J, K and L, respectively. In S1, S2 and W1 mode data, area L corresponds to the maximum fractal value of 2.8 (Table 11.2).

It is interesting to find that the fractal dimension algorithm based probability is able to extract ship wake information in area H with a value of 2.9 (Fig. 11.13). This suggests that the corresponding value of the fractal dimension for different categories allows a multi-fractal characterization of different features in different RADARSAT-1 SAR modes. These results confirm the study of Marghany et al. [206].

In S1, S2 and W1 mode data, it can further be seen that low wind zone in areas M, N and O occur close to the coastline with the maximum fractal values equal to 2.33, 2.34 and 2.5, respectively (Fig. 11.14). Look-alikes occupy narrow areas parallel to the coastline (Fig. 11.14). The wide distribution of dark zone pixels represents the natural slick in low wind areas [197], which is aligned with what could be a current shear or convergence zone. This can be seen clearly in S1 mode data (Fig. 11.14c). Figures 11.14d and 11.14e, however, illustrate the deficiency of oil slick in the pair of S2 mode data. Thus, the fractal algorithm is able to discriminate the look-alike features from the surrounding sea surface features such as current shear (Fig. 11.14d). Figure 11.14d illustrates, however, fewer areas of look-alikes as compared to Fig. 11.14e. The fractal dimension values of look-likes, current shear and ships and their wakes are shown in Figs. 11.13 and 11.14 are approximately similar. Consequently, the non-significant difference has been acquired when we compared between the fractal dimension values of look-alikes, current shear and ship pixels for different acquisition dates (Fig. 11.14). In this context, the t-test value of 16.7 is lower than t-critical with $p < 0.005$ and error of 0.17 (Table 11.3).

Table 11.3. Significant differences between look-alike and low wind zone fractal values in RADARSAT-SAR different mode data.

Data	Statistics Parameters			DF	P < 0.05	Significant
	t-statistics	t-critical	error			
Pair of S2 (1999–2005) and other mode data	1.34	1.69	0.17	400	0.0005	Non-significant

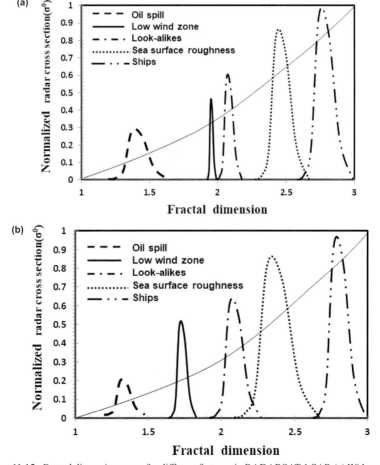

Figure 11.15. Fractal dimension curve for different features in RADARSAT-1 SAR (a) Wide mode (W1), Standard mode (S2) and (c) Standard mode (S1).

Further, in W1 mode data, the sea surface roughness has a fractal value of 3.2 and a normalized radar cross-section of 0.85 whereas the oil spill pixels have fractal dimension values between 1.5 and 1.56 and a normalized radar cross section 0.18, 0.2 and 0.29 in S1, S2 and W1 mode data, respectively (Fig. 11.15). Indeed, the sea surface can be considered as a fractal object. According to Falconer [196], the slope measure of fractal objects corresponds to the complexity of the objects, with the natural implication that the sea surface would have a steady value (Figs. 11.14 and 11.15). There appears to be a reduction in the maximum fractal dimension of the oil slick compared to that of the look-alikes. This could be due to the short spatial extent of the oil spill as shown in the S1 and S2 mode data. The large difference of 1.6 in fractal values occurs between the maximum peaks of look-alike and oil spill, in both modes (Fig. 11.15a). Therefore, the oil spills have a lower normalized radar backscatter cross section in three RADARSAT-1 SAR beam modes as compared to the surrounding sea surface environments, sea surface roughness, low wind zone and look-alikes (Fig. 11.15). This could be attributed to an exponential relationship between fractal surface and normalized radar backscatter cross section (Fig. 11.15b). This result agrees with the study of Bertacca et al. [207].

By contrast, Gado and Redondo [20] found that a box-counting fractal dimension model provided excellent discrimination between oil spills and look-alikes, although they did not consider the backscatter information, which could allow a first robust localization of the oil spills. Benelli and Garzelli [197] used a multi-resolution algorithm, which was based on the fractal geometry for texture analysis. They found that the sea surface is characterized by an approximately steady value of the fractal dimension, whereas oil spills have a different average fractal dimension compared to look-alikes.

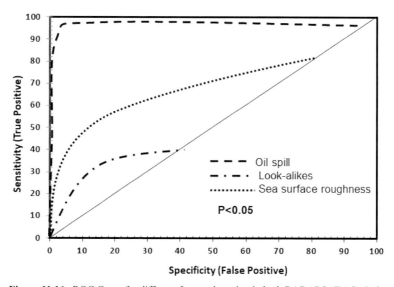

Figure 11.16. ROC Curve for different feature detection in both RADARSAT-1 SAR data.

11.10 How Far Can Fractal Algorithm Detect Oil Spill?

The Receiver-Operator-Characteristics (ROC) Curve in Fig. 11.16 indicates significant differences in the discrimination between oil spill, look-alikes and sea surface roughness pixels. In terms of ROC area, this evidence is provided by an area difference of 15% for the oil spill and 45% for the sea roughness and a *p*-value below 0.005, which confirms the study of Marghany et al. [198]. In fact, the fractal dimension can be viewed as a measure of the scale of the self-similarity of the object. Also, the interference is statistically similar if the scale is reduced, which is similar to the result of Bertacca et al. [207]. This suggests that fractal analysis is a good method to discriminate regions of oil slick from surrounding water features.

Further, Fig. 11.17 shows an exponential relationship between the fractal dimension and the standard deviation of the estimation error for the fractal dimension. The maximum error standard deviation is 0.47, corresponding to the fractal dimension value of 2.9 which is found in S1 mode data. For oil spill detection, the minimum error standard deviation of 0.02 occurs in a region of fractal dimension of 1.49 in W1 mode data. This means that the W1 mode performs better for detection of the oil spill as compared to S1 and S2 mode data. In fact, W1 mode shows a steeper incident angle of 30°, than both S1 and S2 mode data. The offshore wind speed during W1 mode overpass was 4.11 m/s, whereas the offshore wind speed was 7 m/s during the S2 mode overpass. In fact, wind speeds below 6 m/s are appropriate for detection of the oil spill in SAR data [176]. Therefore, for applications that require imaging of the ocean surface, steep incidence angles are preferable as there is a greater contrast of backscatter manifested at the ocean surface.

Good discrimination between oil spill, look-alike, low wind zone and sea surface roughness exists when the error standard deviation is between 0.002 and 0.45 as produced by the implementation of the fractal modified formula. The reason is that the

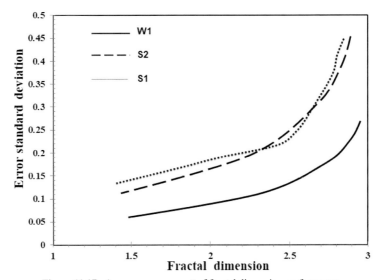

Figure 11.17. Accuracy assessment of fractal dimension performance.

fractal dimension is a measure of the scale of the self-similarity of the object. The low standard deviation error value of 0.002 for a fractal area of 1.49 (Fig. 11.17) dominated by the oil spill is lower than that for the surrounding sea. This is thus an excellent indicator for the validation of the fractal formula modification by implementing a Probability Distribution Function (PDF).

The fractal dimension based on the Probability Distribution Function (PDF) improves the discrimination between oil spill, look-alikes, sea roughness and low wind zones. In fact, involving the PDF formula in the fractal dimension map directly relates the textures at different scales to the fractal dimension. Such a modification of the fractal equation reduces the problems of speckle and sea clutter and assists in the accurate classification of different textures for SAR images.

Previous studies have been concerned with automatically detecting oil spills from SAR images which are based on dark spot feature extraction and classification; Solberg et al. [176]; Mohamed et al. [179]; Samad and Shattri [180]; Benelli and Garzelli [197]; In contrast to this chapter, these studies, however, failed to detect the oil spill spread and discriminate between the current shear features, ship pixels, sea surface roughness and oil spill pixels by using different segmentation algorithms [176,180,184] or the classical fractal formula [197]. Indeed, the different oil spill segmentation approaches in terms of classification accuracy of the oil spill and sea surrounding features are a challenging task, whereas the modification of automatically detecting oil spill's algorithms are required. In conclusion, the modified formula of the fractal box counting dimension has improved the ability to distinguish the oil spill from the surrounding sea surface feature. This has been proved by the capability of the fractal new formula to identify the deficiency of oil slick in pairs of RADARSAT-1 S2 mode data. It can be said that the new approach of the fractal box counting dimension algorithm can be used as an automatic tool for the oil spill, and other sea surface features.

Quantum Cellular Automata Algorithm for Automatic Detection of Oil Spills and Look-Alikes

12.1 Principles of Quantum-dot Cellular Automata

Quantum-dot Cellular Automata (QCA) is a novel and the hypothetically splendid technology for executing computing architectures at the nano-scale. The vital Boolean primeval in QCA is the common gate. This chapter presents a novel design for QCA cells and another possible and unconventional scheme for majority gates for automatic detection of the oil spill. In this view, QCA can be exploited to create common determination computational and memory circuits. QCA, therefore, is presumed to accomplish extraordinary scheme extent, actual great clock frequency and tremendously squat power depletion [209].

Moreover, the interesting key features of QCA is that there is no immobile assembly approach and hence, it should be conceivable to involve such optimization algorithms of genetic algorithms to minimize the number of cells in a design. In this regard, the genetic algorithm can be used for improving hardware performance. In Bonyadi [215] a logic optimization method is introduced and hardware saving using a genetic algorithm is performed [210,214].

The majority gate is the keystone in constructing a block of QCA circuit which can be an alternative of Boolean logic operators (and, or and their complements). In fact, the majority logic signifies and operates digital functions on the foundation of mainstream resolution [209,211,213].

12.2 Quantum Cellular Automata Cell Construction

Quantum Cellular Automata is a novel stratagem construction which is acquiescence to the nanometer scale [216]. The code of quantum cellular automata was first suggested by Lent et al. [217]. So as to gadget a structure that encodes information in the formula of electron position, which it befalls compulsory to paradigm a receptacle in which an electron can be deceived and calculated as there or not there. A quantum dot does just this by creating a constituency of squat prospective enclosed by a loop of extraordinary possibility. In other words, QCA encodes the logic states not as voltage stages, nonetheless as a function of the position of singular electrons [216]. In conventional practice, QCA machinery is constructed on the communication of bi-stable QCA cells created from four quantum dots [209].

A schematic of a basic cell is shown in Fig. 12.1. The cell is charged with two free electrons, which are able to tunnel between adjacent dots. These electrons tend to occupy antipodal sites as a result of their mutual electrostatic repulsion. Thus, there exist two equivalent energetically minimal arrangements of the two electrons in the QCA cell, as shown in Figs. 12.2 and 12.3. These two arrangements are denoted as cell polarization. Binary information is encoded in the charge configuration of the QCA cell to represent logic 1 and 0 [218].

In this understanding, the mathematical expression of cell polarization is given by:

$$P = \frac{(\rho_2 + \rho_4) - (\rho_1 + \rho_3)}{\rho_1 + \rho_2 + \rho_3 + \rho_4} \tag{12.1}$$

Equation 12.1 demonstrates that the electronic charge at the dot ρ_i at i dot. In this view, electrons are precisely confined on sites two and four in which leads to $P = +1$, while electrons on sites one and three yield $P = -1$.

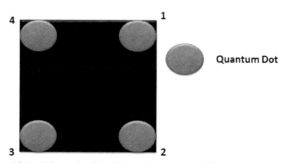

Figure 12.1. Schematic of the basic cell constructed from four quantum dots.

12.3 QCA Adder with Five Gates for Automatic Detection of Oil Spill

For automatic detection of the oil spill in SAR images, the ordinary form of QCA cells is considered. In this sense, QCA can be effectively performed by running the cell from the surrounding sides of the oil spill features which were described earlier in Chapter 10. The route of cell running begins from the bottom, up and left of the

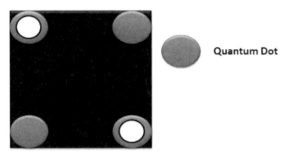

Figure 12.2. Dual bistable states result in cell polarizations of P = +1 (Binary1).

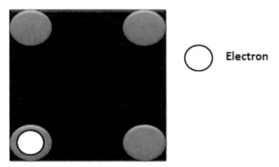

Figure 12.3. Dual bistable states result in cell polarizations of P = −1(Binary 0).

Figure 12.4. The schematic symbol for the majority gate of selected oil spill features.

SAR data. Therefore, we considered five-input majority gates, i.e., complexity (C), entropy (E), decoherence (D), maximum contrast (M), and spreading (S) (Fig. 12.4).

The majority voting logic function can be expressed in terms of the fundamental Boolean operator as:

$$F(C,E,D,M,S) = CED+CEM+CES+CDM+CDS+CMS$$
$$+EDM+EMS+DMS \tag{12.2}$$

This structure makes the majority decision. Therefore, a schematic symbol of a five pins majority gate is revealed in Fig. 12.4. Let us consider a three-input AND gate and also a three-input OR gate using the following formula:

$$F(C,E,D,0,0) = CED \tag{12.3}$$

$$F(C,E,1,1,S) = C+E+S \tag{12.4}$$

Let us then take 1-bit full adder, which involves input features of the oil spill and its surrounding environment and output of either oil spill or look-alikes. To this end, the input features are C and E and the carry bit is the decoherence D. Figure 12.5 reveals the design QCA for automatic detection of the oil spill in the SAR data. In this design three majority gates and two inverters are considered. In this regard, the QCA inverter presents dual standard cells in a diagonal orientation, which are geometrically similar to dual rotated cells in a horizontal orientation. For this reason, standard cells in a diagonal orientation tend to align in opposite polarization directions as in the inverter chain [209,216].

Following Zhang et al. [218], a procedure for reducing the quantity of majority gates is required for computing three fluctuating Boolean functions to facilitate the conversion of the sum-of-products expression into QCA majority logic. In this regard, these standard functions are used to convert the sum-of-products expression of majority logic (Fig. 12.5).

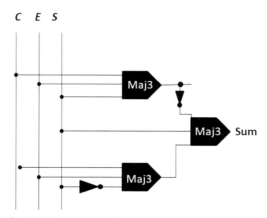

Figure 12.5. One-bit QCA adder with five gates for automatic detection of the oil spill.

12.4 Cellular automata for Automatic Detection of Oil Spill

Let consider I_{SAR} as the set of SAR feature integers. Therefore, 2-D cellular space of SAR image can be presented as is a 4-tuple, i.e. $(I_{SAR} \times I_{SAR}, \psi, N, T)$. In these circumstances, $I_{SAR} \times I_{SAR}$ a set of the Cartesian product of two integer sets of SAR images, while ψ is a set of cellular states. Moreover, T is the local transition function from ψ^n to ψ as N is the type of neighborhood. Therefore, the relevant neighborhood function is a function from $I_{SAR} \times I_{SAR}$ to $2^{I_{SAR} \times I_{SAR}}$, which is mathematically formulated as:

$$g(F) = \{F + \delta 1, ..., F + \delta_n\} \tag{12.5}$$

where F is the selected oil spill features, which $F \in I_{SAR} \times I_{SAR}$ and $\delta_i (i = 1, 2, 3, ..., n) \in I_{SAR} \times I_{SAR}$ is fixed. For 2-D von Neumann neighborhood, the neighborhood state function of the central cell F is expressed as:

$$H^t(F) = \left(\psi^t(F + (0,0))\right), \left(\psi^t(F + (0,1))\right), \left(\psi^t(F + (1,0))\right),$$
$$\left(\psi^t(F + (0,-1))\right), \left(\psi^t(F + (-1,0))\right) \tag{12.6}$$

where $\left(\psi^{t}\left(F+(0,0)\right)\right)$ is the current state of the central cell of the oil spill selected features $\left(\psi^{t}\left(F+(0,1)\right)\right)$ and $\left(\psi^{t}\left(F+(0,-1)\right)\right)$ are presented by the north and (south) cells, respectively. In contrast $\left(\psi^{t}\left(F+(1,0)\right)\right)$ and $\left(\psi^{t}\left(F+(-1,0)\right)\right)$ are presented by the east and west cells, respectively. Conversely, let us correlate the neighborhood state of a cell F at time t to the cellular state of that cell at time $t+1$ as:

$$f\left(H^{t}\left(F\right)\right) = \psi^{t+1}(F) \tag{12.7}$$

Equation 12.7 reveals that the function f is denoted to the 2-D CA rule and is regularly specified in the form of a state table, specifying all possible pairs of the form $\left(H^{t}\left(F\right)\right), \psi^{t+1}\left(F\right)\right)$. In this context, Fig. 12.6 reveals 2-D SAR image and 2-D CA.

The local transition function $T: \Sigma^{N} \rightarrow H_{\Sigma}$ where H_{ψ} is the Hilbert space spanned by the cell states ψ. This model can be viewed as a direct quantization of a CA where the set of possible configurations of the CA is extended to include all linear superpositions of the classical cell configurations, and the local transition function can map the cell configurations of a given neighborhood to a quantum state (Fig. 12.7). One cell is labeled "accept" cell. The quiescent state is used to allow only a finite number of states to be active and renders the pixels effectively finite. This is crucial to avoid an infinite product of unitaries and, thus, to obtain a well-defined QCA.

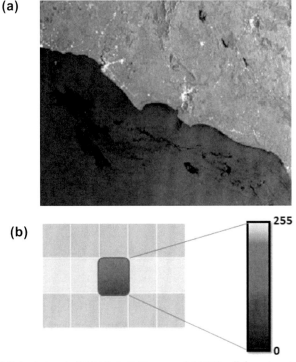

Figure 12.6. Cellular Automata (a) SAR digital image and (b) 2-D cellular automata with 256 states.

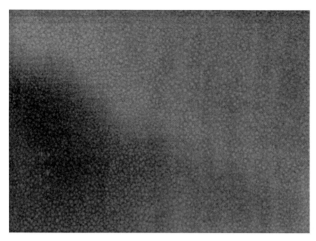

Figure 12.7. SAR image cell configuration by QCA.

In the Schr"odinger picture of quantum mechanics, the state of a system at some time t is described by a state vector $|\psi_t\rangle$ in a Hilbert space H $|\psi_t\rangle$ (Fig. 12.8). The state vector evolves unitarily,

$$|\psi_{t+1}\rangle = U|\psi_t\rangle. \tag{12.8}$$

here U is a unitary operator, i.e., $UU^\dagger = 1$, with the complex conjugate $U^\dagger$ and the identity matrix 1. If $\{|\psi_i\rangle\}$ is a computational basis of the Hilbert space H$|\psi_t\rangle$ any state $|\psi\rangle \in H$ can be written as a superposition $\sum_{|\phi_i\rangle} c_i|\phi_i\rangle$ with coefficients $c_i \in \mathbb{C}$ and $\sum_i c_i c_i^* = 1$.

The automatic detection of the oil spill in a SAR image is implemented by using Block-partitioned Quantum Cellular Automata (QCA) (Fig. 12.9). In this regard, Block-partitioned QCA is a 4-tuple. To this end, given a system with nearest-neighbor

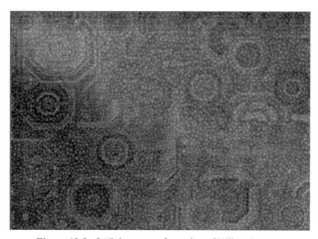

Figure 12.8. SAR image configuration of Hilbert Space.

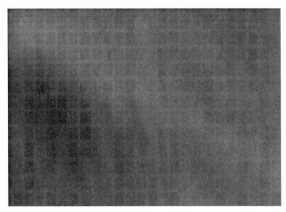

Figure 12.9. Block-partitioned quantum cellular automata of SAR image.

interactions, the simplest unitary QCA rule has radius $r = 1$ describing a unitary operator applied over a three-cell neighborhood $j-1,j$, and $j+1$ as:

$$M\left(u_{00}, u_{01}, u_{10}, u_{11}\right) = |00\rangle\langle00| \otimes u_{00} + |01\rangle\langle01| \otimes u_{01} +$$
$$|10\rangle\langle10| \otimes u_{10} + \qquad\qquad (12.9)$$
$$|11\rangle\langle11| \otimes u_{11},$$

Equation 12.9 demonstrates that the term $|ol\rangle\langle ol| \otimes u_{ol}$ presents an update of the of qubit at site j with the unitary u_{ol} if the qubit at the site $j + 1$ is in the state $|l\rangle$. In this understanding, states $|l\rangle$ and $|o\rangle$ are presented look-alikes and oil spill respectively. M transforms with its own 2-site conversion. Thus, partitioning of SAR image for the oil spill and look-alikes is presented by updating simultaneously entirely even qubits with rule M prior to updating entirely odd qubits with rule M. Periodic boundaries are expected. Nonetheless, by statement ability of the end qubits simulation of a block-partitioned QCA by a QCA with boundaries can be achieved (Fig. 12.10). These boundaries are clearly presented in Fig. 12.10.

Figure 12.10. Qubit block-partitioned for SAR image.

Let us consider a nearest-neighbor 1-dimensional Block-partitioned QCA. In the density operator formalism, each quantum system ρ is specified by the probability distribution function as:

$$\rho = \sum_i p_i |\psi\rangle\langle\psi|$$ (12.10)

where p is the probability of the quantum state ψ which presented in Equation 12.8. to this end, A completely positive map $S(\rho)$ applied to state ρ is represented by a set of Krauss operators K_μ, which are positive operators that sum up to the identity $\sum_\mu K_\mu K_\mu^\dagger = 1$. The mathematical description of the automatic detection of the oil spill

$$S_j^{ol}(\rho) = |ol\rangle\langle ol| \otimes \sum_\mu K_\mu^{ol} \rho K_\mu^{ol\dagger} \otimes |ol\rangle\langle ol|$$ (12.11)

The implementation of such a block-partitioned nonunitary QCA is suggested in the configuration of a pixel of even order created with an irregular array of dual distinguishable oil spill features CSCSCS, which are comprehensively well pronounced and interact via the Ising interaction [219].

With the intention of characterizing the dynamics of entanglement, it would be helpful to have a single parameter that quantifies the amount of multi-particle entanglement contained in a state at any given time step. A virtuous quantity of entanglement should apprehend the nonlocal nature of the quantum correlations of the spins and therefore, should be a function of the state that is non increasing, on average, under local operations and classical communication. In this regard, the amount of multi-spin entanglement with a function of pure states of n qubits acquaint with [220]:

$$R(|\psi\rangle) = 2\left(1 - n^{-1}\sum_{j=0}^{n-1} Tr[\rho_j^2]\right).$$ (12.12)

Equation 12.12 reveals the linearly R relates to the purity of the single qubits averaged over the pixels and satisfies two important properties. First $0 \le R(|\psi\rangle) \le 1$, where $R(|\psi\rangle) = 0$ iff $|\psi\rangle$ presents look-alikes, $R(|\psi\rangle) = 1$ for oil spill entangled states. However, the second $R(|\psi\rangle)$ is invariant under local unitary U_j [220].

12.5 Explored SAR Images

In this study, RADARSAT-2 SAR data acquired by RADARSAT-2 operating in ScanSAR Narrow single beam mode (Table 12.1) on April 27th, 2010; May 1st 2010; and May 3rd, 2010 are investigated for oil spill detection in the Gulf of Mexico. The satellite is equipped with Synthetic Aperture Radar (SAR) with multiple polarization modes, including a fully polarimetric mode in which HH, HV, VV, and VH polarized data are acquired. Its highest resolution is 1 m in Spotlight mode (3 m in Ultra Fine mode) with a 100 m positional accuracy requirement. In the ScanSAR narrow beam mode, the SAR has a nominal swath width of 500 km and an imaging resolution of 100 m. Finally, the wind speed data are retrieved from the studies of NOAA OR&R, [221]; Lynn et al. [222]; and Nan et al. [223]. The validation of oil spill occurrences in RADARSAT-2 SAR data was validated using the ground information collected by

Table 12.1. RADARSAT-2 Characteristics.

RADARSAT-2 Characteristics	Values
Bandwidth (MHZ)	100
Polarization	HH
Nominal Resolution (m)	50 x 50
Center Frequency (GHz)	5,405
Swath Width (km)	300–500
Repeat Cycle (days)	12
Looks	4
Incident angle (°)	20–46

NOAA/NESDIS [224]. The repeat cycle of ScansAR narrow beam mode is 12 days with a number of 4 looks. Its incident angle is ranged between 20°–46° (Table 12.1).

12.6 Oil Spills in RADARSAT-2 SAR Data

An oil platform located 70 km from the coast of Louisiana sank on Thursday, April 22, 2010, in the Gulf of Mexico, spilling oil into the sea (Fig. 12.11). In the RADARSAT-2 SAR Scan Narrow Beam (SCNB) data that were acquired at the time, the rapid growth of the oil slick footprint from April 27, 2010, to May 5, 2010, can be clearly seen (Fig. 12.12). Figure 12.12 shows that oil slick and sheen extended across 19,112 square miles (49,500 km²) of the Gulf. In addition, it is worth noting that the oil slick spun in a counter-clockwise direction. This behavior is attributed to the influence of the Gulf Stream. Nevertheless, the RADARSAT-2 SAR data did not indicate that the oil-slick footprint coincided with the loop current in the Gulf of Mexico. Consistent with Zangari [225], the oil slick was the one caused by the disconnection of the loop current in the Gulf of Mexico.

Figure 12.11. Oil spill disaster in the Gulf of Mexico (Free to use).

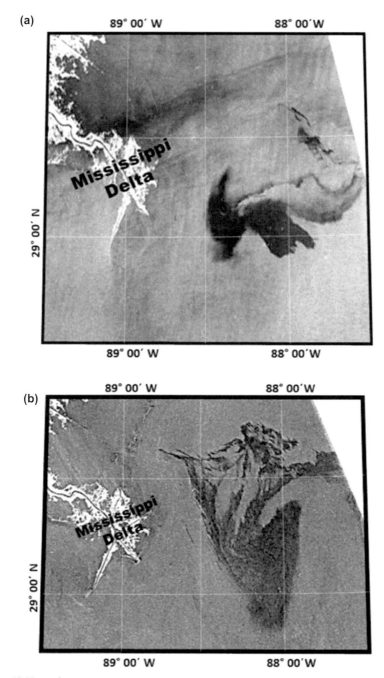

Figure 12.12 contd. ...

... Figure 12.12 contd.

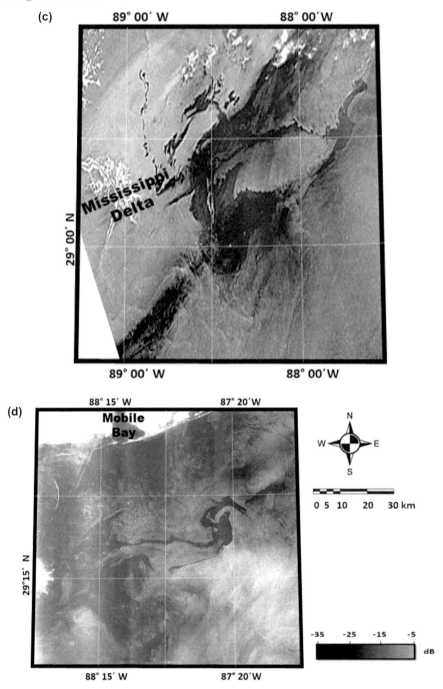

Figure 12.12. RADARSAT-2 SAR Scan Narrow beam SCNB data in (a) April 27th, (b) May 1st, (c) May 3rd, and (d) May 5th 2010.

It is clear that the oil spill, which covered an area of greater than 150 km² in the Gulf of Mexico, had a darker tone than both the surrounding water and some boats in the area (Fig. 12.13). The data represent some regions that are located at the farthest extent of the area that is probed by RADARSAT-2 SAR Scan Narrow Beam (SCNB) with a revisit period. Figure 12.12 shows the variation in the average backscatter intensity along with the oil slick footprint. The average backscatter intensity was dumped by –30 dB to –25 dB and decreased over time as the oil slick footprint gradually increased (Fig. 12.12). By contrast, ship footprints are characterized by a maximum backscatter of –5 dB (Figs. 12.12 and 12.13) because the SCNB mode has nominal near and far resolutions of 7 m.

Clearly, the oil spill is portrayed in SCNB mode by shallower incidence angle ranges between 36° to 46° and the maximum wind speed of 10 m/s (Fig. 12.14). Clearly, the backscatter increases with rising wind speed and decreases with increasing incidence angle. This study confirms the work of Cheng et al. [226]; Caruso et al. [227] and Garcia-Pineda et al. [228]. However, the concept detection of oil spills in SAR images requires moderate wind speeds, not exceeding 6 m/s. Nevertheless, the SCNB mode beam date acquired on May 3rd and 5th (Figs. 12.12c and 12.12d) respectively, can detect oil spill under extremely high wind speed of 10 m/s (Fig. 12.14) which caused the intense turbulence on the sea surface (Figs. 12.12c and 12.12d). This could be due to the extreme amount of oil leakage (approximately 60,000 barrels per day) which cannot be dissipated by such extreme wind and turbulent conditions [228].

The SCNB mode provides images of very wide swaths in a single pass of the satellite with single linear HH co-polarization and a pixel spacing of 25 range x 25 azimuth (m). The SCNB mode provides coverage over the shallow incident angle range of 31° to 47°. This data source was selected because of its large swath (300 km), acceptable pixel size, spacing (25 m), high temporal resolution (2/3 side-lap pass

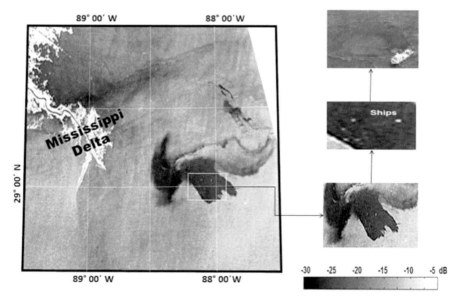

Figure 12.13. Backscatter of ship pixels which are identified by a white box.

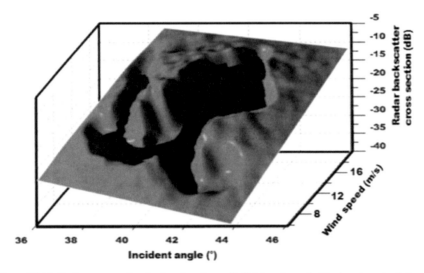

Figure 12.14. Radar cross-section backscatter along oil slick locations, incident angle and wind speed distribution during the date of acquisitions.

within 7 days) and relatively low volume of data. The C-band and shallow incidence angle (31°–47°) data have been found to be suitable for the identification of oil slick footprints. The sensitivity of the SAR backscatter measurements to the water surface roughness created by wind-induced ripples can be reduced by using HH polarization and a large incidence angle. Furthermore, the RADARSAT-1 SAR, when operating in the ScanSAR Narrow mode with a swath width that exceeds 300 km, is a promising tool for marine oil pollution detection [228,229].

12.7 Automatic Detection of Oil Spill by Quantum Cellular Automata

Noticeably, the QCA algorithm is able to discriminate dark oil spill pixels from the surrounding environment. In the algorithm's output, look-alike features, low wind zones, rough patches on the sea surface, and land are indicated by white-colored regions, whereas oil spill pixels are marked in black (Fig. 12.15). Figure 12.15 presents the QCA results in which 100% of the oil slick pixels in the test set were correctly classified. Consequently, Fig. 12.16 confirms the ability of the QCA for automatic detection and discrimination of small oil spill sizes from the large oil spill. It is interesting to note that small oil spill sizes are located out of the area of large ones or at the edge of large oil spill areas (Fig. 12.16). This is excellent evidence that a crossover process is commenced by searching and matching small cell of oil spills (Fig. 12.15b). In fact, the completely positive map $S(\rho)$ has reconstructed the full prototype of large oil spill cells (Figs. 12.15 and 12.16).

In both Figs. 12.15 and 12.16, QCA is able to identify automatically the edges of oil spills. In fact, QCA has sharped the edge of the oil spill after speckle noise reductions. In addition, QCA has the potential to preserve the edges of small oil spills and automatically identified them from the large oil spills and sea surface wave

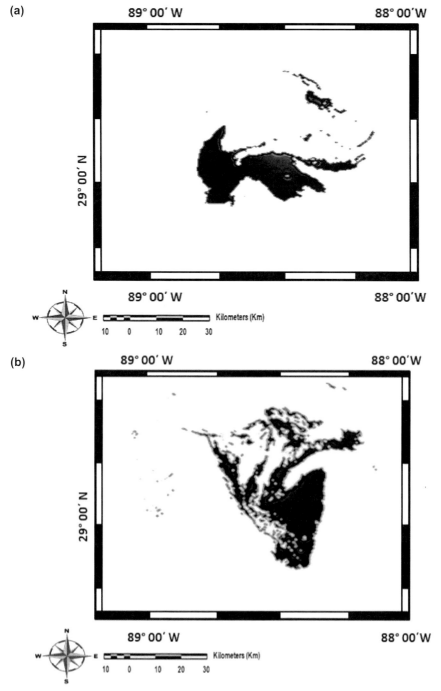

Figure 12.15 contd. ...

... Figure 12.15 contd.

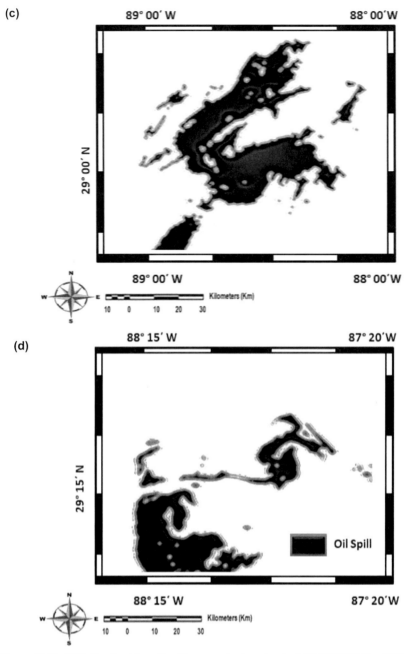

Figure 12.15. QCA for algorithm oil spill automatic detection in RADARSAT-2 ScanSAR Narrow during (a) April 27th, (b) May 1st, (c) May 3rd, and (d) May 5th 2010.

and current interactions. In other words, an edge result of the oil spill exhibits the superb quality with one pixel wide and the edge has no break. In fact, the QCA cell is positioned diagonally to maximize the distance between the input and output QCA cells, which reduces the speckle noise impacts and produces the inversion of the five-input majority gate as a function of oil spill characteristics. In this respect, the QCA obviously delivers the promising result.

12.8 Accuracy of QCA for Automatic Detection of Oil Spill in SAR Data

The automatic detection of the oil spill and look-alikes by using QCA algorithm is appraised based upon the number of False-Positive (FP) and False-Negative (FN) pixels in the regions demarcated by the five major gates. An FN pixel is defined as a pixel in the reference region that is not present within the segmented region. An FP pixel is defined as a pixel outside the reference region that was included in the pectoral region segmented. The false-positive and false-negative are respectively, given by:

$$N_c\text{-FP=TP} \tag{12.13}$$

$$N_u\text{-FN= TN} \tag{12.14}$$

where N_c and N_u present number of changed and unchanged pixels which are detected by QCA. In this chapter, TP and TN present true-positive and true-negative, which represent change and unchanged pixels correctly. In this understanding, the percentage of the correct oil spill and look-alike classifications is given by:

$$\%P_{cc} = \frac{\text{TP+TN}}{N} \tag{12.15}$$

However, the number of pixels N has a large number which makes the similarity between the oil spill and look-alikes difficult to be achieved. To overcome such issue overall error ε_o is determined by:

$$\varepsilon_o = F_P + F_N \tag{12.16}$$

Then the Kappa coefficient k_c is used to determine the overall accuracy as a function of *TP* and *TN*. This can be mathematically expressed by:

$$k_c = \frac{P_{cc} - \left[\left(\text{TP+F}_p\right)N_c + \left(F_N + \text{TN}\right)N_u\right]}{1 - \left[\left(\text{TP+F}_p\right)N_c + \left(F_N + \text{TN}\right)N_u\right]} \tag{12.17}$$

Image classification by QCA is effected kappa coefficient (k_c). In other words, high performance of QCA can be achieved by high-value of k_c. In this regard, k_c is ranged between 0 and 1. Therefore, the high performance of QCA is close to 1 with *less* performance time T *(sec)*.

Table 12.2 shows the rate percentage values of the FP, FN, the error of pixel detections for the result of the QCA algorithm with four RADARSAT-2 SAR images.

Table 12.2. Rate percentage of FP and FN pixels for the results of QCA.

Features	% P_{cc}	k_c	T(sec)
Oil spill	94.23	0.92	72
Look-alikes	92.53	0.89	83
Large oil spill	92.32	0.82	75
Small oil spill	93.55	0.90	63

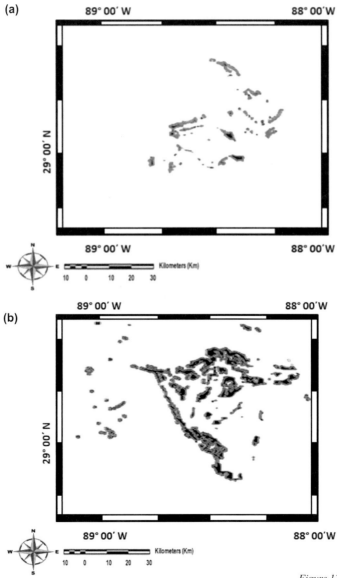

Figure 12.16 contd. ...

... *Figure 12.16 contd.*

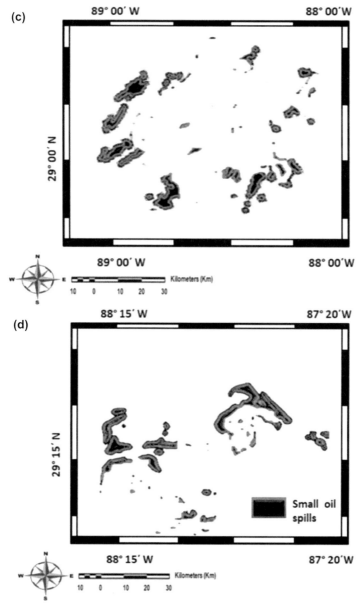

Figure 12.16. QCA algorithm for detection of small oil spills in RADARSAT-2 ScanSAR Narrow during (a) April 27th, (b) May 1st, (c) May 3rd, and (d) May 5th 2010.

It is interesting to find that oil spill is detected by high rate of P_{cc} with 94.23% and high k_c value of 0.92 than look-alikes, large and small oil spill. This could be due to the interaction of the lookalikes with oil spills. However, small oil spill pixels are detected faster than oil spill and look-alikes within 63 seconds. In fact, the small oil

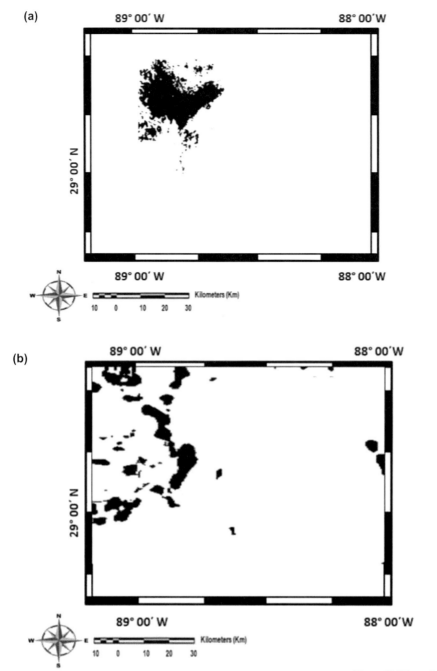

Figure 12.17 contd. ...

... Figure 12.17 contd.

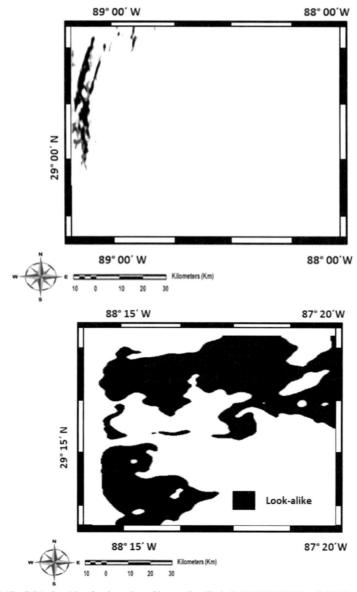

Figure 12.17. QCA algorithm for detection of large oil spills in RADARSAT-2 ScanSAR Narrow during (a) April 27th, (b) May 1st, (c) May 3rd, and (d) May 5th 2010.

spill is a small stream of a larger one so it can be faster detected by QCA. As QCA algorithm is also able to detect automatically the look-alikes (Fig. 12.17). It is worth mentioning that the locations of look-alikes are determined clearly by using QCA. In fact, QCA algorithm based on the five majority gates are able to discriminate between the physical properties of both look-alikes and oil spills. In this manner, look-alikes are located close to the coastline, which indicates wind shelter zone (Fig. 12.17a), or stream bloom (Figs. 12.17b and 12.17c) from the Mississipi Delta. However, Fig. 12.17d suggests the interaction between look-alikes and oil spill, which confirms the result of Figs. 12.15 and 12.16. This also agrees with Table 12.2. In this regard, the proposed QCA-based algorithm is one of the promising methods in automatic detection of the oil spill and look-alikes.

These findings confirm the results of the study of Marghany [229]. Nevertheless, the work of Marghany [229] cannot track small oil spill pixels from the large ones. It suggests that QCA is an excellent way to separate small oil spill pixels from the large ones and also from other features such as look-alikes and sea surface roughness.

The power of the QCA lies in the block-partitioned nonunitary. In this view, the dual distinguishable features of oil spills are considered through the irregular array as it delivers accurate automatic detection of an oil spill from the surrounding environment.

As a result, the quiescent state allows only a requested state of the oil spill and look-alikes to be active which assist in avoiding an infinite of unitaries. In this understanding, the QCA algorithm can detect independently both oil spill and look-alikes state. Thus, the new state of large oil spills are continuously generated based on the dissimilarities between $|\psi\rangle = 0$, which identifies look-alikes, and $|\psi\rangle = 1$, which presents an oil spill entangled state. Therefore, the measurement reveals a one or a zero—the choice randomly determined from the probabilities of the superposition. Measuring sets the value of the qubit. If we measure the qubit and get a one, checking again will also result in a one, which is talent validation of QCA algorithm. In the point of view of quantum computing, if we consider oil spill state and look-alike state as the inner product $|ol\rangle\langle ol|$ which allows us to describe the probability of occurrence of look-alike $|l\rangle$ state in the state of the oil spill $\langle o|$, which has been proven by Fig. 12.17d.

In addition, entanglement can be shared in different approaches by different subsets (parties) of the spins in the SAR image, there is no single function that describes multi-partite entanglement, which produces a more refined oil spill pattern, i.e., small and large oil spill pixels by despeckling and maintaining the morphology of the features of the oil spill pattern by virtue of the multi-partite entanglement that is used to implement the oil spill pixel classification. Indeed, QCA algorithm evolution can consider a configuration of spins equipped with a product state to a number of different entangled states. In this understanding, it automatically selects a morphological pattern for the small oil spill pixels than large ones that are similar to the requested small and large spill arch types.

12.9 Why QCA is able to Detect Oil Spill Automatically?

Previous studies were concerned with automatic detection of oil spilling from SAR images, which is based on the dark spot feature extraction and classification using Artificial Neural Networks (ANN) [172,173], fractal [185], Texture Algorithm [163], and Local Probability Maximization (LPM) [230]. Therefore, QCA provided excellent automatic detection of small oil spills as compared to previous studies. In addition, the QCA algorithm is also able to isolate look-alikes from the surrounding environment, while none of the previous studies achieved this task. In fact, automatic segmentation of the oil spill problem is expressed as an optimization problem of the oil spill boundaries and QCA algorithm proficiently detect the global maximum in a search space of SAR images and resolves the problem of parameter selection in SAR image segmentation through five major gates. Further, the QCA algorithm is implemented in SAR image segmentation to modify oil spill selected parameters in existing segmentation algorithms and pixel-level segmentation. In this manner, quantum computers do work in a superposition but saying that "they can store a large amount of data" is very wrong. If a qubit array is accessed like a digital memory, it will randomly spit out only one of the values in the superposition, and the array will collapse to this same value, the superposition is destroyed. The superposition is also very unstable, almost anything can change it, thermal noise, electromagnetic fields, mechanical vibrations, etc., it usually lasts for a few milliseconds.

Moreover, the RADARSAT-2 data are encoded into the qubits, which are put into a superposition state that holds the probabilities of finding either oil spill or look-alikes. Since the search is completely random, both oil spill and look-alike features are represented with equal probability.

Finally, the QCA algorithm has the ability to determine the optimal number of regions of even small oil spill segmentation or to choose some features such as the size of the analysis window or some heuristic thresholds.

In general, RADARSAT-2 SAR data deliver the extreme level of flexibility compared to other SAR sensors. Indeed, a couple of days for revisiting cycle of SCNB have provided both a wide area coverage and high-resolution imaging of oil slick detection and monitoring Macondo spill in the Gulf of Mexico. In this regard, for marine environmental monitoring, the ScanSAR modes with resolutions of 20–60 m and swath widths of 100–300 km can provide an ideal trade-off between spatial resolution and areal coverage. This resolution is adequate for supporting oil spill automatic detection. Further, SCNB shows the potential even to identify small oil spill pixels from the surrounding large oil spill pixels by using such a QCA algorithm.

Quantum Multiobjective Algorithm for Automatic Detection of Oil Spill Spreading from Full Polarimetric SAR Data

The segmentation problem of an oil spill or look-alikes using partition clustering is viewed as a combinatorial optimization problem. But the prevailing optimal approaches, e.g., Conventional Genetic Algorithms (CGAs), are often time-consuming, and their convergences speed is slow and easy to trap in local optimal value.

Conventional algorithms ensue single phase at a time since a turning machine can only be in a single position a time. Consistent with quantum physics, the elementary oil spill and ocean wave-particles of nature are not in one fixed state at any moment, but can occupy several phases at ones, in what is called superposition. When these particles are scattered in the SAR images, they decohere into one state. In Chapter 12, it was proven that a quantum algorithm is able to distinguish automatically between look-alikes and oil spills.

Quantum computing is recognized for its origins of quantum mechanics, for instance, uncertainty, superposition, interference and implicit parallelism. These advantages make it have a superior variety and the superior steadiness between the investigation and the utilization than conventional evolutionary algorithms. Various scientists have been acquainted with quantum codes into evolutionary computing. For instance, Han and Kim [231], suggested a quantum-inspired evolutionary algorithm for combinatorial optimization as a function of the perception and codes of quantum computing. In this regard, Jiao and Li et al. [232] deliberated a quantum-inspired immune clonal algorithm for global optimization.

Automatic detection of the oil spill in SAR image is dominated by defecting of the single-objective clustering algorithms in addition to the uncertainty of darkness patch segmentation as an oil spill. In this view, Chapter 12 cannot answer what is the level of oil spill spreading. It is an intention of this chapter to solve this analytical problem. There is no doubt that the oil spill has incredible shapes than other materials floating on the sea surface. Can a quantum multiobjective genetic algorithm determine the level of oil spill spreading for a fully polarimetric SAR data?

13.1 Principles of Fully Polarimetric SAR Images

Consider the fully-polarized SAR image S_{xy}, in which oil spill pixel backscattering properties can be identified by a backscattering matrix as:

$$S = \begin{pmatrix} S_{HH} & S_{HV} \\ S_{HV} & S_{VV} \end{pmatrix}$$ (13.1)

where S_{xy} is the complex backscattering efficiency, with x denoting the received wave polarization and y indicating the transmitted wave polarization. **Therefore**, the radar transmits an only linear combination of horizontal and vertical ($\pi/4$) or circularly (CTLR, DCP) polarized signal and linearly (CTLR, $\pi/4$) or circularly (DCP) receives both horizontal and vertical polarizations in In the compact polarimetric SAR modes. In this view, the 2-D quantity vector $\vec{K}$ is the projection of the complete backscattering matrix on the transmit polarization state. Hence, the quantity vector $\vec{K}$ of three foremost compact polarimetric SAR modes can be delineated as:

$$\vec{k}_{\pi/4} = [S_{HH} + S_{HV} \ \ S_{VV} + S_{HV}]^T / \sqrt{2}$$ (13.2)

$$\vec{k}_{CTLR} = [S_{HH} - iS_{HV} \ \ iS_{VV} + S_{HV}]^T / \sqrt{2}$$ (13.3)

$$\vec{k}_{DCP} = [S_{HH} - S_{VV} \ \ i2S_{HV} \ \ i(S_{HH} + S_{VV}]^T / 2$$ (13.4)

Following Souyris et al. [253], the magnitude of linear coherence and the cross polarization ratio can be correlated with parameter γ as:

$$\gamma = \frac{1 - |\rho_{HHVV}|}{\langle |S_{HV}|^2 \rangle} \times \langle |S_{HH}|^2 \rangle + \langle |S_{VV}|^2 \rangle$$ (13.5)

Equation 13.5 demonstrates that the probability of polarimetric SAR oil-spill sorting leans on greatly dissimilar polarimetric mechanisms for clean water surface and oil-covered [254]. In this context, based on different polarimetric scattering behaviors, mineral oil and biogenic slicks can be better distinguished: for the oil-covered area, Bragg scattering mechanism is largely suppressed and high polarimetric entropy can be witnessed, while in case of a biogenic slick, Bragg scattering is still dominant, but with a lower intensity. In this understanding, cross and co-polarization Ratio, which is the power ratio between HH and VV or HV and HH/VV channels can be obtained through dual-pol systems. In tilted Bragg scattering model adopted by Minchew et al. [254], cross and co-Polarization ratio are only a function of dielectric constant and incidence angle. Nonetheless, definite circumstances are continuously beyond obscured and further investigations are greatly required. Co-polarization ratio was also proved to be possible to discriminate slicks from look-alike features associated with low-wind conditions and surface current effects [255].

Polarimetric SAR decomposition parameters entropy (E) and average alpha angle ($\bar{\alpha}$) can be depleted to investigate the scattering patterns of oil slicks rivaled to the slick-free ocean. In this regard, E measures the randomness of the scattering mechanisms and ($\bar{\alpha}$) portrays the scattering mechanism. For oil-covered areas, damping is strong, the correlation between co-polarization channels is low and thus E and ($\bar{\alpha}$) are high. In this view, non-Bragg scattering dominates are considered. Nonetheless, for the clean sea surfaces, the relationship between co-polarization channels is high, and E and ($\bar{\alpha}$) are low, indicating that surface Bragg scattering dominates [252].

13.2 Quantum Computing

In the quantum computer, a qubit is used rather than the classical bit. Quantum systems are described by a wave function y that exists in a Hilbert space. The Hilbert space has a set of states, ψ, that form a basis, and the system is described by a quantum state y, which is in a linear superposition of the basis states ψ, and in the general case, the coefficients $\mathbb{C}$ may be complex. A qubit is a vector in two-dimensional complex vector space [233]. In this space, a vector has two components and the projections of the vector on the basis of the vector space are a complex number. In this regard, the quantum state can be described as:

$$|\psi> = \alpha \,|\,0> + \beta\,|\,1> \tag{13.6}$$

$$|\dot{\alpha}^2| + |\beta^2| = 1 \tag{13.7}$$

here α and β are the complex numbers and $|0>$ and $|1>$ are the dual orthonormal bases for the dual dimensional vector space [233–236]. It can be shown that the qubit is a linear superposition of the bases state, by changing the values of the coefficients.

$$|\psi> = \cos\theta\,/\,2\,|\,0> + \exp(i\phi)\sin\theta\,/\,2\,|\,1> \tag{13.8}$$

Equation 13.8 demonstrates that the qubit can be represented as a vector r from the origin to a point of the three-dimensional sphere with a radius of one, is known as the Bloch sphere (Fig. 13.1).

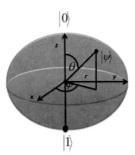

Figure 13.1. Bloch sphere.

Qubits behave like bits. They can be on or off, but in effect, they can be on and off at the same time. On the contrary, the traditional bit does not have a feature corresponding to this one, where it takes only either state 0 or state 1. Nevertheless, the qubit can take any value represented by the vector *r* on the Bloch sphere (Fig. 13.2).

The quantum system is said to collapse when the projection on one of the basis is achieved. That is also called decoherence or the measures. For instance, in the circumstance of the projection of $|\psi>$ on the $|0>$ basis then it should be $|\psi>=\alpha|0>$. Therefore, $|\dot{\alpha}^2|$ is the probability of the qubit to collapse on the state $|0>$. Therefore, in Quantum Multiobjective Evolutionary (QME), a novel Q-bit chromosome representation is adopted based on the concept and principles of quantum computing, such as Q-bits and linear superposition [231]. The characteristic of the representation is that any linear superposition of solutions can be represented. The smallest unit of information stored in a dual-state quantum computer is called a quantum bit (Q-bit), which may be in the "1" state, or in the "0" state, or in any superposition of the two. The state of a Q-bit can be represented by Equation 13.8.

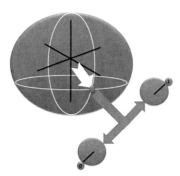

Figure 13.2. Qubit representation.

13.3 Quantum Machine Learning

In the quantum machine learning two phases are involved: (i) learning phase and (ii) pattern recall phase.

13.3.1 Learning Phase

The quantum machine learning is based on the Grover algorithm, which is based on constructing a coherent quantum system. In this view, m presents the pattern and it is known as the pattern-sorting process. Let us assume a set $D=\{\psi(P)\}$ of polarimetric SAR features which has *m* patterns of the oil spill and look-alike features as a training set, the aim is to produce $|\psi\rangle$ such that

$$|\psi(P)\rangle = \sum_{m\in P}\psi(P)|P\rangle \tag{13.9}$$

here *P* represents the different patterns in the training set **D**. Moreover, *P* can be represented in a binary format b_1, b_2 and b_3. Using Grover algorithms, *n* qubits are required to represent *P* patterns, in addition to $n+1$ qubits as required control qubits in the learning algorithms. That provides the sum of *2n+1* qubits, which are necessitated to characterizes *m* patterns. It also delivers the sum of 2n+1 registers in implanting the quantum network. A state $|\psi\rangle$ *can* be constructed using a polynomial number (*n* and *m*) of elementary operations on one, two or three qubits. That means the $|\psi\rangle$ will be constructed recursively using three operations [236–239]. In this view, the mathematical description of the coherence state of the polarimetric SAR features $|\psi_{PolSAR}\rangle$ is given as:

$$|\psi_{PolSAR}\rangle = |x_1 x_2 \ldots\ldots, x_n, R_1 R_2, \ldots, R_{n-1}, C_1 C_2\rangle \tag{13.10}$$

Equation 13.5 indicates the restoring P patterns as $x_1 x_2 \ldots\ldots, x_n$, with the total number *n*. In addition, $R_1 R_2, \ldots, R_{n-1}$ as well as $C_1 C_2$ are controlled register patterns. Grover algorithm, therefore, involves three operations, which deliver the coherent state of the polarimetric SAR features $|\psi_{PolSAR}\rangle$.

The first operation involves the matrix generation *M* which is described by:

$$M = \begin{bmatrix} 1 & 0 & 0 & 0 \\ 0 & 1 & 0 & 0 \\ 0 & 0 & \sqrt{\dfrac{p-1}{p}} & \dfrac{\psi(m)}{\sqrt{p}} \\ 0 & 0 & \dfrac{\psi(m)}{\sqrt{p}} & \sqrt{\dfrac{p-1}{p}} \end{bmatrix} \tag{13.11}$$

M must be achieved under the circumstance of $1 \le P \le m$ and $1 <= P <= m$. Then the second operation is presented by the flip transformation ψ^0 which is described by:

$$\psi^0 = \begin{bmatrix} \psi & 0 \\ 0 & I \end{bmatrix} \tag{13.12}$$

This equation is used if there are dual qubits, i.e, one for the oil spill and another is for look-alikes and it is required that one of them control the flip transformation on the other one. In other words, let us assume the state of the oil spill in polarimetric SAR image is presented by $|\psi_{Oil} \ge 1|0\rangle$, while the state of look-alikes is identified by $|\psi_{Look-alike} \ge \beta|0\rangle + \alpha|1\rangle$. In this understanding, the pattern state detection of oil spill using the M matrix is given by:

$$\Psi^0 |oil\rangle\langle look| = \begin{bmatrix} 0 & 1 & 0 & 0 & 1 & \beta \\ 1 & 0 & 0 & 0 & 1 & \alpha \\ 0 & 0 & 0 & 1 & 0 & \beta \\ 0 & 0 & 1 & 0 & 0 & \alpha \end{bmatrix} \tag{13.13}$$

Then the third operation is based on the transformation Ψ^{00}, which is a 3-qubit operation. It flips the state of one qubit if and only if the other two are in the $|00\rangle$

states. In this regard, 3-qubits $|\psi_1 \geq 1|0>$ and $|\psi_2 \geq 1|0> \ |\psi_3 \geq \beta|0> + \alpha|1>$ are assumed and represented by the levels of oil spill spreading, i.e., thick, medium and light, respectively. This can be mathematically described as:

$$\Psi^{00}|\psi_1\psi_2\psi_3\rangle = \begin{bmatrix} \alpha \\ \beta \\ 0 \\ 0 \\ 0 \\ 0 \\ 0 \\ 0 \end{bmatrix} \tag{13.14}$$

13.3.2 Pattern Phase Recall

A pattern examination is achieved post of the learning phase $\Psi^{00}|\psi_1\psi_2\psi_3\rangle$ which has all the patterns of oil spill levels in the SAR data. In fact, the main hypotheses are recalling the pattern from the required pattern basis at state $|\psi_{SAR}\rangle$ [237–240]. Let us recall the oil spill pattern 01 from the 2-qubits. $|\psi_{oil}\rangle$ should collapse on the desired state after repeating the algorithm for $\left(\dfrac{3.14}{4}\right) \times 2$ one time iteration. For instance, $|\psi_{oil}\rangle$ can be achieved as $-\left[\sqrt{3}\right]^{-1}|01\rangle + \left[\sqrt{3}\right]^{-1}|10\rangle - \left[\sqrt{3}\right]^{-1}|11\rangle$. Then the algorithm computes the average with $|01\rangle\left[\sqrt{3}\right]^{-1}(0,-1,1,1)$, which is equal to $0.25\sqrt{3}$. Conversely, the entire quantum set is rotated around the average, for instance, $|\psi > 0.5\sqrt{3}(1,2,-1,-1)$. Finally, the probability of the algorithm to collapse on the oil spill state is, for instance, ¾= 75%. In this circumstance, the higher the number of qubits the higher the percentage of collapsing at accurate patterns of the oil spill.

13.4 Quantum Multiobjective Evolutionary Algorithm (QMEA)

Consistent with Wang et al. [241], evolutionary computing with Q-bit representation has a better characteristic of population diversity than other representation since it can represent a linear superposition of state probabilities. The only one above Q-bit individual is enough to represent eight states, while in binary representation, at least eight strings (000), (00 1), (010), (011), (100), (101), (110) and (111) are needed (Fig. 13.3). Further, in QMEA crossover and mutation operations are used to maintain the diversity of the population, while evolutionary operation that quantum gates operate on the probability amplitudes of basic quantum states is applied to maintain the diversity of the population in the Quantum Genetic Algorithm (QGA) and Quantum Evolutionary Algorithm (QEA).

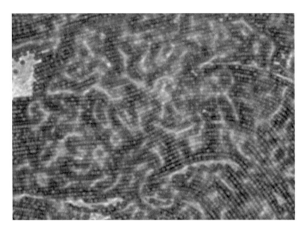

Figure 13.3. The initial generation of qubit of SAR images.

13.5 Generation of Qubit Populations

Let us assume that $Q(t) = \{q_1^t, q_2^t, ..., q_n^t\}$ (Fig. 13.4) is the individual qubit population, where $q_1^t = \begin{Bmatrix} \alpha_{i1}^t, \alpha_{i2}^t, ..., \alpha_{im}^t \\ \beta_{i1}^t, \beta_{i2}^t, ..., \beta_{im}^t \end{Bmatrix}$ $(i = 1, 2, ..., n)$ is a qubit individual, n is the dimension of population and m is the span of the individual. In the preliminary stage, the qubit individuals are with a similar probability of $\dfrac{1}{\sqrt{3}}$.

α_{i1}^t	....	....	....	α_m^t
β_1^t	....	....	....	β_m^t

Figure 13.4. An array of one qubit q_i^t populations.

13.6 Generation of Oil Spill Population Pattern

Let us assume that the classical binary oil spill population $O(t) = \{o_1^t, o_2^t, ..., o_n^t\}$ where $(i = 1, 2, ..., n)$ with length m is a binary string, which is observed from q_i^t. In this view, the random number of oil spill population polarimetric features is generated with the constraint of $N_O \in [0,1]$. In this circumstance, the random number N_o is validated as an oil spill when $N_O > |\alpha_i^t|^2$ in which the corresponding bit in o_n^t equal to 1. On the contrary, if o_n^t equals to 0, the polarimetric SAR features do not present an oil spill. Therefore, the clustering point center of the oil spill is presented by bit 1 in the binary individual.

13.6.1 Quantum Fitness

The oil spill footprint boundaries must be coded into the qubit chromosome form. In this problem, the qubit chromosome q_m consists of a number of qubits where every qubit corresponds to a coefficient in the north-order surface fitting polynomial as given by:

$$f(i, j)_q = q_0 + q_1 i + q_2 j + q_3 i^2 + q_4 ij + q_5 j^2 + \ldots + q_m j^n \tag{13.15}$$

The fitness is then evaluated using the connectedness metric. In other words, the degree of the qubit chromosomes that are signed from surrounding points in polarimetric SAR image and clustered as oil spill can be mathematically described as:

$$conn(q^t) = \sum_{i=1}^{m} \left(\sum_{j=1}^{L} (q_{i,j}^t) \right) \tag{13.16}$$

Equation 13.16 delivers the approach of identification of oil spill clusters from the surrounding environment of a number of qubit chromosomes L with m size. In this circumstance the oil spill identified by every 10 qubit chromosomes for every point nearest point i and its neighboring j-th. Under the circumstance of $q_{i,j}^t = j^{-1}$ else $q_{i,j}^t = 0$, i.e., absent of oil spill occurrence in polarimetric SAR image and feature belongs to look-alikes.

13.6.2 Quantum Mutation

In QEA, the qubit representation can be used as a mutation operator (Fig. 13.5). Directed by the current best individual, the quantum mutation is completed through the quantum rotation gate.

$$G(\theta) = \begin{bmatrix} \cos\theta & -\sin(\theta) \\ \sin\theta & \cos\theta \end{bmatrix} \tag{13.17}$$

where θ is the rotation angle of the quantum rotation gate (Fig. 13.6) and is described as:

$$\theta = k \times f(\alpha_i, \beta_i) \tag{13.18}$$

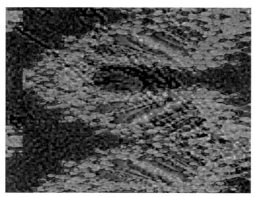

Figure 13.5. Quantum mutation of polarimetric SAR image for detection of the oil spill.

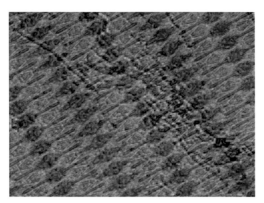

Figure 13.6. Quantum rotation gate for detection of the oil spill in SAR data.

here k is a coefficient and the value of k has an impact on the speed of convergence. The value k must be chosen reasonably. If k is too big, the search grid of the algorithm is large and the solutions may diverge or have **a** premature convergence to a local optimum, and if it is too little, search grid of the algorithm is also little and the algorithm may be in a stagnant state. In addition, $f(\alpha_i, \beta_i)$ determines the search direction of convergence to a global optimum [242]. The quantum gate is the keystone to determine the false alarm clustering of oil spills or true clustering of oil spills. Figure 13.6 demonstrates the initial recognition of the oil spill as dark patches by using a quantum rotation gate. Table 13.1 delivers the look-up table to determine the best solution of the quantum chromosome of oil spills.

The updated procedures can be given by:

$$q_i^{t+1} = G(t) \times q_j^t \tag{13.19}$$

The probability amplitude of the qubit is denoted by q_i of **an** individual at *t-th* generation as t is evolutionary generation at the quantum gate $G(t)$.

Table 13.1. Lookup Table of $f(\alpha_i, \beta_i)$.

$\|\alpha_1\| * \|\beta_1\| > 0$	$\|\alpha_2\| * \|\beta_2\| > 0$	$f(\alpha_i, \beta_i)$	
		$\arctan\left(\|\beta_1\| / \|\alpha_1\|\right)$	$\arctan\left(\|\beta_2\| / \|\alpha_2\|\right)$
		(Best Solution)	(Current Solution)
True	True	+1	−1
True	False	−1	+1
False	True	−1	+1
False	False	+1	−1

13.7 Quantum Non-dominate Sort and Elitism (QNSGA-II)

Here a brief description of QNSGA-II relevant to this study is presented. QNSGA-II is the second version of the famous "*Non-dominated Sorting Genetic Algorithm*" based on the work of Prof. Kalyanmoy Deb for solving *non-convex* and *non-smooth* single and multi-objective optimization problems. Its main features are: (i) a sorting non-dominated procedure of qubit chromosome $Q(q_N^t)$ where all the individual are sorted according to the level of non-domination; (ii) it implements elitism which stores all non-dominated solutions, and hence enhancing convergence properties; (iii) it adopts a suitable automatic mechanics based on the crowding distance in order to guarantee diversity and spread of solutions; and (iv) constraints are implemented using a modified definition of dominance without the use of penalty functions [243].

Perhaps, there is not one best solution in the case of multiple objectives that exists. Therefore, there are a set of solutions which are superior to the rest of solutions in the search space when all objectives are considered but are inferior to other solutions in the space in one or more objectives. These solutions are known as Pareto-optimal solutions or nondominated solutions [245–251].

The efficiency of QNSGA lies in the way multiple objectives are reduced to dummy fitness function using nondominated sorting procedures. Consequently, QNSGA can solve practically any number of objectives. In this regard, this algorithm can handle both minimization and maximization problems [244].

In order to sort a population of size N for qubit chromosome $Q(q_1^t),\dots\dots,Q(q_N^t)$ according to the level of non-domination, each solution m must be compared with every other solution in the population to find if it is dominated. This requires comparisons $O(Q(q_m))_N$ for each solution, where m is the number of quantum qubit chromosome of different pixels belongs to oil spills, look-alikes and sea roughness, and low wind zones. The initialized population N $Q(q_1^t),\dots\dots,Q(q_N^t)$ is sorted based on the level of non-domination. Let $S_{Q(q_1^t)}$ is every generated quantum multiobjective solution to determine accurate level of domination. In this regard, the fast sort algorithm was given by Deb et al. [243] can be explored in oil spill automatic detection as follows:

for each individual $Q(q_1^t)$ in main population P do the following:

Initialize $S_{Q(q_1^t)} = \Phi\left(Q\left(q_N^t\right)\right)$. This set Φ would include all the individuals of $Q\left(q_N^t\right)$ which is being dominated by $Q(q_1)$.

Initialize $n_{Q(q_1)} = 0$. This would be the number of individuals that dominate $Q\left(q_N^t\right)$, i.e., no individuals dominate $Q(q_1)$ then $Q(q_1)$ belongs to the first front; set rank for an individual $Q(q_1)$ to one, i.e., $Q(q_1)_{rank} = 1$.

for each individual m in $P(Q(q_1))$

if $Q(q_1)$ dominated m then
- add m to the set $\Phi\left(Q\left(q_N^t\right)\right)$, i.e., $\Phi\left(Q\left(q_N^t\right)\right) = \Phi\left(Q\left(q_N^t\right)\right) \cup \{m\}$

    *else if m dominates $Q(q_1)$ then
- the increment for domination counter for $Q(q_1)$, i.e., $n_{Q(q_1)} = n_{Q(q_1)} + 1$

Let the first front set $F_1\left(Q\left(q_N^t\right)\right)$ and then update by adding $Q(q_1)$ to front 1, i.e.,

$$F_1\left(Q\left(q_N^t\right)\right) = F_1\left(Q\left(q_N^t\right)\right) \cup \{Q(q_1)\}$$

Initialize the front counter to one. $i = 1$

Then $F_i\left(Q\left(q_N^t\right)\right) \neq \Phi\left(Q\left(q_N^t\right)\right)$

Let $Q(q_1) \neq \Phi\left(Q\left(q_N^t\right)\right)$. The set for sorting the individuals for $(i+1)^{th}$ the front

for each individual $Q(q_1)$ in front $F_i\left(Q\left(q_N^t\right)\right)$

For every individual m in $S_{Q(q_1^t)}$ (is the set of individuals dominated by $Q(q_1)$)

- $n_{Q(q_1)} = n_{Q(q_1)} - 1$, decrement the domination count for individual m.

- if $n_{Q(q_1)} = 0$ then none of the individuals in the subsequent fronts would dominate m. Hence set $Q(q_1)_{rank} = i+1$. Update the set $Q(q_i^t)$ with individual m, i.e., $Q(q_i^t) = Q(q_i^t) \cup m$.
  - increment the front by one.
  - For oil spill, now the set $Q(q_i^t)$ is the next front and hence $F_i(Q(q_i^t)) = Q(q_i^t)$.

13.8 Quantum Pareto Optimal Solution

In this chapter, the multi-objective of polarimetric SAR features for oil spill detection as discussed earlier is considered. Let $P_{max}(Q(q_i^t)$ is quantum qubit chromosomes of probability amplitude of oil spill footprint in polarimetric SAR features which should satisfy:

1. turbulent flow pixels $(P_{max}(Q(q_i^t)))$: the variation of maximum quantum qubit chromosomes of probability which contain the oil spill quantum qubit chromosomes, i.e., $P_{max}(Q(q_i^t)) = max\left\{P_1(Q(q_1^t)),....., P_k(Q(q_k^t))\right\}$, where $P_jQ(q_j^t)$ denotes probability occurrence of the oil spill in Pareto Front j, $\forall j = 1,2,...,k$.

2. total probability of quantum qubit chromosomes of the oil spill $(\sum P_i(Q(q_i^t)_i)$: the sum of quantum qubit chromosomes in each row and column in SAR data.

Pareto optimal solutions are applied to retain the discrimination of oil spill diversity and its surrounding environment. The level of convergence to the Pareto front $P_f\left(P_i(Q(q_i^t))\right)$ is calculated due to the Euclid distance, as follows:

$$P_f\left(P_i(Q(q_i^t))\right) = \sum_{\omega=1}^{\omega}\sqrt{\sum_{m=1}^{N}\min_{u=1,U}\left(P_{um}(Q(q_{um}^t)) - a_{\omega m}\right)^2} \tag{13.20}$$

Equation 13.20 determines the performance of the Pareto front for automatic detection of the oil spill and its spreading using QEA. In this view, N presents the number of optimized criteria for automatic detection of oil spill spreading in polarimetric SAR images and um is the number of Pareto front points (Fig. 13.7), which are delivered from the probability occurrence of quantum qubit chromosomes of the oil spill The Pareto front algorithm generates the tested points $a_{\omega m}$.

Figure 13.7. Pareto front curves generated with quantum qubit chromosomes in polarimetric SAR data.

13.9 Automatic Detection of Oil Spill in Full Polarimetric SAR

The SAR images are acquired by UAVSAR (uninhabited aerial vehicle synthetic aperture radar) throughout the DWH (Deepwater Horizon) oil spill disaster in the Gulf of Mexico on June 23, 2010. UAVSAR is a fully-polarimetric L-band SAR. They have a center frequency of 1.2575 GHz. Moreover, they have dual multi-look, i.e., 3 and 12 looks in the range and azimuth directions, respectively. UAVSAR data have an incident angle ranges between 22° to 65°. UAVSAR data have a 5 m slant range resolution and 7.2 m azimuth resolution. In this section, two UAVSAR data were acquired from two adjacent, overlapping flight tracks, which covered the core oil spills in the Gulf of Mexico. The two flight lines are gulfco_14010_10054_100_100623 (Fig. 13.8a) and gulfco_32010_10054_101_100623 Fig. (13.8b), respectively [252]. It is clear that the oil spill pixels in both images have damping of normalized radar cross-section NRSC of -35dB along the oil spill pixels.

UAVSAR instrument has a lower noise floor range between -35dB to -53dB, which allow for better oil spill imagine. In fact, the instrument noise floor is grasped only at the far edge of the swath for the HV backscatter from oil-covered. This can help to quantify the radar cross-section of water with an L-band radar, even with oil damping the surface waves. Specific quantities are operated for parameter settings, i.e., quantum antibody population, encoding length, cloning scale and maximum iteration number. In this regard, we set N equals 100, encoding length L equals 20, under the circumstance of 100 iteration number to ensure an accurate result. Figure 13.13 delivers the results, which are obtained by quantum NSGA-II (QNSGA-II). QNSGA-II can discriminate automatically between Bragg scattering and non-Bragg scattering. The dynamic fluctuation between both Bragg and non-Bragg scattering is accurately determined. Moreover, QNSGA-II can deliver several clusters in UAVSAR, for instance, surrounding sea surface, light oil spill and thick oil spill (Fig. 13.9).

(a)   (b)

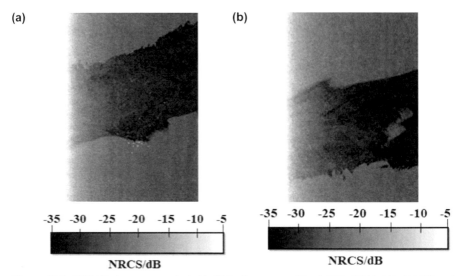

Figure 13.8. UVSAR data are acquired (a) flight lines are gulfco_14010_10054_100_100623 and (b) gulfco_32010_10054_101_100623.

Figure 13.9. Oil spill discrimination in UAVSAR data using QNSGA-II.

Subsequently, we reset the N equals 80, encoding length L equals 20, and increase the iteration number to 100, we acquire sharp oil spill morphology features (Fig. 13.10). In addition, the clear discrimination of thin oil spill and thick oil spill are also shown in Fig. 13.10.

The increment of the number of parameters leads to the accurate clusters of the oil slick. In other words, these oil spill clusters have revealed the variety of the oil spill's characteristics, from thicker, concentrated emulsions to minimal oil contamination (Fig. 13.10). In fact, UAVSAR characterizes an oil spill by detecting variations in the roughness of its surface and, for thick slicks, changes in the electrical conductivity of its surface layer. In this regard, UAVSAR "sees" an oil spill at sea as a smoother

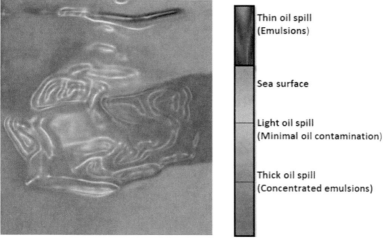

Figure 13.10. Thin and thick oil spill discriminations in UAVSAR image using QNSGAII.

Color version at the end of the book

(radar-dark) area against the rougher (radar-bright) ocean surface because most of the radar energy that hits the smooth surface is deflected away from the radar antenna. UAVSAR's high sensitivity and other capabilities enabled the team to separate thick and thin oil for the first time using a radar system. This result is validated by another UAVSAR data which was acquired on June 23, 2010, shows the southern Louisiana coastline, covering the area around Grande Isle and the entrance to Barataria Bay (Fig. 13.11a). It is better to implement VV polarization as it is sensitive to scattering from the sea surfaces. QNSGAII delivers the clear spreading levels an of the oil spill, which involves thick, medium and thin (Fig. 13.11b). It is clear that the oil spreads from the open ocean into wetland areas past the coastline.

The colors in the image reflect the three different UAVSAR radar polarizations: HH (horizontal transmit, horizontal receive) is colored red; VV (vertical transmit, vertical receive) is colored blue; and HV (horizontal transmit, vertical receive) is colored green. Like a pair of Polaroid sunglasses, these images are sensitive to different parts of the radar signal that is reflected back from the Earth's surface. The HV polarization is sensitive to multiple scattering that typically occurs in vegetation. VV polarization is sensitive to scattering from surfaces—this gives a bluish tint to water and non-vegetated soil. Finally, HH polarization is sensitive to structures and vertical tree trunks—this gives some urban areas and vegetated regions a reddish tint. Dark areas on the water surface are caused by something that smooths the surface and damps wave activity, such as oil.

Consistent with Leifer et al. [257] the high spatial resolution of the UAVSAR enabled the characterization of very near-shore oiling of vegetation in the marshlands affected by the DWH oil spill. The radar backscatter intensity for the different polarizations in the Bay Jimmy area to the northeast of Barataria Bay, Louisiana, showed evidence of surface oil slicks (Fig. 13.10). These slicks, which a ground crew observed, were mainly identified as sheen on the day of the UAVSAR overflight,

(a)

(b)

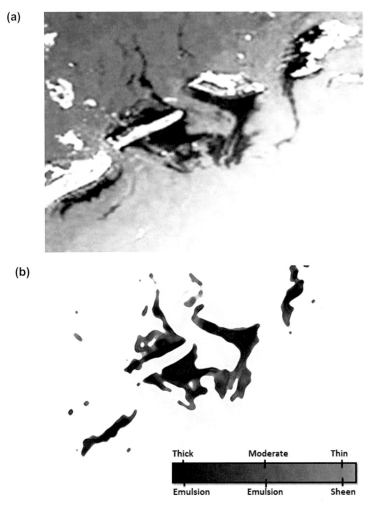

Figure 13.11. False color UAVSAR L-band data (a) June 23, 2010 and (b) QNSGA-II for automatic detection of oil spill spreading.

and show up as radar-dark areas within the bay. Barataria Bay winds during the two days of data acquisitions were 2.5–5.0 m s^{-1} and 2.5–3.5 m s^{-1} from the southeast. UAVSAR returns from unslicked water in the Gulf and Barataria Bay were similar (Fig. 13.11) and much higher than oil slick returns in either the main slick or Barataria Bay at all incidence angles. Furthermore, oil slick returns from the main slick were significantly higher than oil sheen returns in Barataria Bay, indicating sensitivity in the UAVSAR data to varying oil properties including thickness and weathering. It is agreed that the thin sheen ranges from 0.1 to 1.0 μm, while the thick emulsion layers is more than 50 μm [257].

The advantages of global search in quantum immune and spectral embedding in UAVSAR data are combined in QNSGA-II. QNSGA-II is able to find the optimal solutions more frequent with the maximum clustering accuracies in 100 runs. In fact, QNSGA-II is operated in both optimal solution and stability. It appears that the QNSGA-II output result is superior, which spectacles more of the oil clusters and less false alarms than the raw UAVSAR images and QEA (Figs. 13.8 and 13.9). It is easy to realize an unbroken oil slick in the QNSGA-II image. Moreover, QNSGA-II can distinguish between emulsions, Minimal oil contamination, and concentrated emulsions. It is clear that QNSGA-II is able to identify the boundary current impact the spreading of oil spill levels, i.e., thick and light (Fig. 13.11). In fact, the thick oil spill is dominated by a high level of the decoherence than the surrounding sea rough surface.

13.10   Applications of QNSGA-II to other Satellite Polarimetric SAR Sensors

QNSGA-II is implemented also on satellite polarimetric SAR sensors which involved RADARSAT-2 quad-polarization (VV and VH) SAR image of oil slicks in the Gulf of Mexico on May 8, 2010 (Fig. 13.12). Therefore, the quad-pol mode on Radarsat-2 has only a spatial coverage of 25 km × 25 km (nominal swath width of 25 km) [256]. The oil spill is dominated by damping backscatter of -35dB (Fig. 13.12)

Figure 13.13 depicts the reveal results by QNSGA-II for quad-polarization data. It is clearly noticed that QNSGA-II can automatically detect oil spill from its surrounding area. In addition, it delivers two clusters for the oil spill, which includes thicker, concentrated emulsions and minimal oil contamination.

Subsequently, QNSGA-II can also detect only the thick oil spill from **the** coherency of quad-polarization (Fig. 13.14). In fact, the magnitude of the correlation coefficient between HH + VV and HH − VV are useful for suppressing lookalikes [255]. Moreover, quad-polarization SAR images are susceptible to noise. Pauli decomposition has the

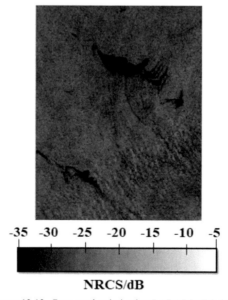

Figure 13.12. Raw quad-polarization RADARSAT-2 data.

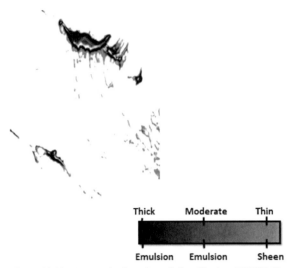

Figure 13.13. Automatic detection of oil spill using QNSGA-II.

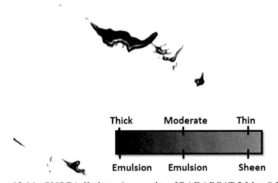

Figure 13.14. QNSGA-II clustering results of RADARSAT-2 May 8 2010.

advantages of anti-interference and general high adaptability [253–257]. In general, the Pauli decomposition images are clearer than original quad-polarization SAR images. The image preprocessing stages are as follows: (i) the original quad-polarization SAR data are decomposed by Pauli, and (ii) the obtained Pauli decomposition images are filtered by 3 x 3 Boxcar filtering. However, QNSGA-II is able to optimize the impact of noise in addition to providing automatic clustering and automatic detection of the oil spill in quad-polarization SAR data.

Generally, the QNSGA-II is able to deliver the same results with single RADARSAT-2 data which were used before in Chapter 11. In this case, Fig. 13.15 shows the variety of clusters which include land, sea surface and two patterns of an oil spill, which are highly concentrated emulsions and low minimal oil contamination. Figures 13.16 and 13.17 confirm that QNSGA-II can deliver the same results for both single SAR polarization and quad-polarization data.

In this e regard, QNSGA-II is also implemented to TerraSAR-X strip map (HH/VV) data over the Kerch Strait in November 2007 (13.18a). It is interesting to note

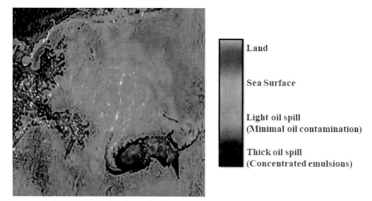

Figure 13.15. QNSGA-II clustering for RADARSAT-2 acquired on April 27th, 2010.

Color version at the end of the book

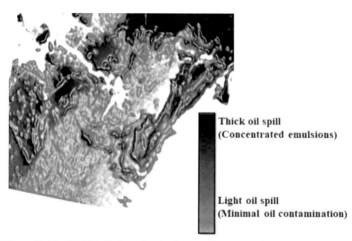

Figure 13.16. QNSGA-II clustering for RADARSAT-2 acquired on May 3rd, 2010.

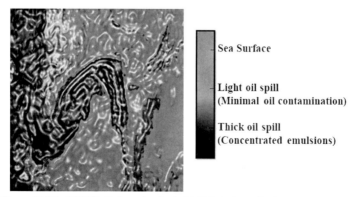

Figure 13.17. QNSGA-II clustering for RADARSAT-2 acquired on May 5th, 2010.

Color version at the end of the book

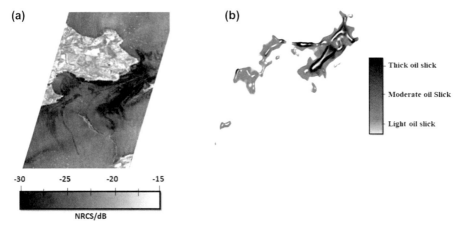

Figure 13.18. Oil spill clusters which are obtained from (a) TerraSAR-X strip map by (b) QNSGA-II.

that QNSGA-II produces three clusters, which are a thick oil spill, medium oil spill and light oil spill too (13.18b). This indicates the level of oil spill spreading under the dynamic action of both wind and sea surface hydrodynamic impacts in the Kerch Strait, Russia. Moreover, QNSGA-II delivers an excellent description of the oil spill edge features in SAR satellite data. The edge of concentrated emulsions appear thicker than minimal oil contamination (Fig. 13.17).

The recent data which verified oil spill in the Malacca Straits is Sentinel-1A (Fig. 13.19a). Sentinel-1A data were acquired on January 6th, 2019 with beam mode IW and VV and VH polarization. It is interesting to find that the ships have the highest backscatter of -5dB while the oil spill with a wind speed of 6 m s^{-1} has the lowest backscatter of -35 dB. In this sense, QNSGA-II is shown to greatly distinguish between

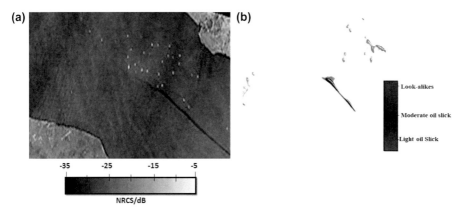

Figure 13.19. The oil spill and look-alikes clusters are acquired from (a) Sentinel-1A data by (b) QNSGA-II.

the oil spill and look-alikes (Fig. 13.19b). Moreover, QNSGA-II is able to determine the degree of oil spill spreading from thick to moderate state.

QNSGA-II provided unexpected precious clusters of oil spill characteristics, which could be related to thickness, from polarimetric SAR data and dual polarization data too. The ROC curves of QNSGA-II with different sensors, which are presented with a higher true positive rate (TPR). In terms of the ROC area, an oil spill can be clustered by 99% among the different SAR sensors by using QNSGA-II (Fig. 13.20).

QNSGA-II provides, a strong elitist approach with mechanisms to sustain the diversity efficiently using non-dominated sorting using multiple observations of Q-bit individuals, which allow a local search in the vicinity of the non-dominated solutions. Also, maintaining the best Q-bit individuals in every generation can avoid the possibility of losing high-quality individuals [265].

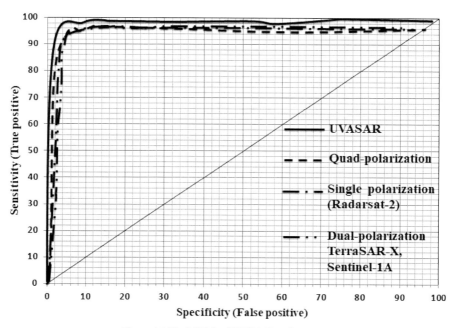

Figure 13.20. ROC for QNSGA-II performance.

13.11 Quantum Decoherence Theory and QNSGA-II

The oil-covered water is considered as decoherence zone while the surrounding sea surface is coherence. In quantum mechanics, quantum decoherence is the loss of coherence or assembling of the phase angles between the constituents of oil-covered water and surrounding sea surface in a quantum superposition. One significance of this quantum decoherence of oil-covered water is conventional or probabilistically addictive behavior. Quantum decoherence offers the attendance of wave function collapse, i.e., Bragg scattering and non-Bragg scattering. In other words, quantum decoherence is considered the decline of the physical characteristics of sea roughness backscatter into a single possibility of oil-covered water, which is assumed as decoherence pixels in SAR images. In this understanding, decoherence speculation of oil-covered water in SAR data is the mechanism, which determines the location of the quantum-classical boundary, i.e., Bagg scattering. In this view, decoherence occurs when oil-covered water interacts with its surrounding environment in a delectrical irreversible approach. In this circumstance, thick oil-covered water with the lowest delectrical than its surroundings prevents Bragg and non-Bragg scattering in the quantum superposition of the entire SAR scene's wavefunction from interfering with each other. This is why the dumping of backscatter occurs in the oil-covered area than its surroundings. However, decoherence does not spawn definite wave function collapse. It only delivers a description of the accomplishment of wave function collapse, as the quantum nature of the oil spill in the SAR data " disintegrates" from its surrounding environment. That is, components of the wavefunction are isolated from a coherent sea surface and attain phases from their instantaneous surroundings. In this circumstance, a total superposition of the global or universal wavefunction immobile exists (and remains coherent at the global level of the sea roughness), nonetheless its vital providence remains as an interpretation concern along the oil-covered water. In this regard, QNSGA-II can play a vital role to determine the lost information in the oil spill pixels. In addition, QNSGA-II can solve the computational gridlock problems of conventional spectral clustering algorithms on large-scale SAR image segmentation. In fact, QNSGA-II signifies the oil spill representative pixels and boundaries through an operative encoding technique. In this understanding, the QNSGA-II optimizes the selection of representative oil spill pixels, while the calculated affinity function for the clustering oil spill and its surrounding environments diminishes the computation quantity to a huge SAR data used. Subsequently, QNSGA-II can count the huge amount of eigenvalues with a value of one, which leads to the precise determination of cluster number across the SAR data. In other words, the attendance of qubit distribution in this algorithm allows for excellent separation between different clusters. It is clear that the boundaries between different oil spill clusters are well identified. Indeed, the largest eigenvalue is set as one, and the other eigenvalues diminish steadily, while the breach between the second and third eigenvalues, is prevalent, which reveals that the three or two-classes of the oil spill can be easily separated.

In the earlier Chapters 11 and 12 , it was described that QNSGA-II is accomplished better than that quantum segmentation algorithm, particularly in clustering the oil spill pattern. In fact, QNSGA-II is able to preserve the morphology of oil spill footprint boundaries, i.e., spreading. Moreover, the conventional segmentation approaches, especially for full polarimetry SAR for instance, Entropy E, Anisotropy A, Alpha angle α and Co-Polarized Phase Difference (CPD) arc not able to cluster the degree of oil spill variations from thick to thin due hydrodynamic diffusion. However, co- and cross-polarization ratio, the degree of Polarization, Entropy H, Alpha angle α, Anisotropy A [254–257] under the control of certain incidence angle range, can still distinguish oil spill from its surrounding environment. In this sense, QNSGA-II is accurately performed better than the co-polarized phase difference, Conformity coefficient, Co-pol correlation coefficient for oil spill automatic detection and clustering.

13.12 Pareto Optimization Role in QNSGA-II

QNSGA-II algorithm considers both local examination and global search into account, which could handle the optimal solution with higher precision compared to conventional evolutionary algorithms [258]. In this context, the QNSGA-II algorithm can be used for function optimization and multiuser detection of code division multiple access [258–260]. Quantum computing and the developed algorithm can be implemented with a genetic algorithm to improve its accuracy [265]. Finally, the quantum image processing is required for advanced SAR image segmentation and processing.

QNSGA-II can accurately identify the morphological boundary of oil spill spreading from thick to light and assigned by different segmentation layer in multi SAR data. In fact, QNSGA-II provides a set of compromised solutions called Pareto optimal solution since no single solution can optimize each of the objectives separately. The decision maker is provided with a set of Pareto optimal solutions in order to choose a solution based on the decision maker's criteria. This sort of QNSGA-II solution technique is called a posteriori method since the decision is taken after the search is over. In this context, the quantum Pareto-optimization approach does not require any a priori preference decision between the conflict of the oil spill, look-alike, land and surrounding sea footprint boundaries. Further, quantum Pareto-optimal points have formed Pareto-front as shown in Figs. 13.21 to 13.23 in the multi-objective function of the full polarimetric SAR features space.

The nondominated solution of different algorithms is presented from Figs. 13.21 to 13.23. It is clear that the solution of QNSGA-II (Fig. 13.21) is much better than

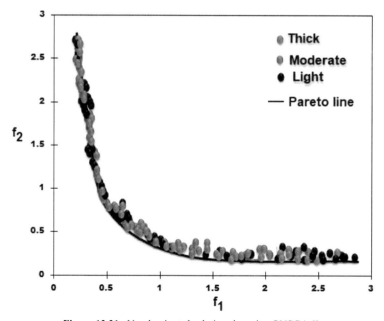

Figure 13.21. Nondominated solutions by using QNSGA-II.

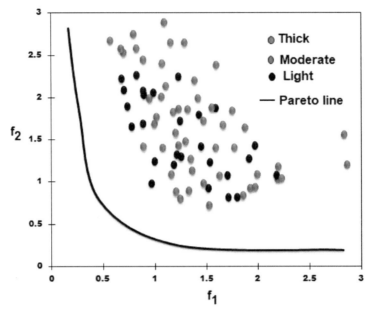

Figure 13.22. Nondominated solutions by using Entropy.

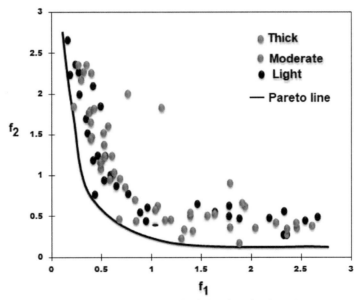

Figure 13.23. Nondominated solutions by using Q-MEA.

entropy (Fig. 13.22) and Q-MEA (Fig. 13.23). Further, Entropy solution is far from the real Pareto front while the solution of Q-MEA is gathered around the center of the Pareto front. Under this circumstance, Q-MEA tends to concentrate on one part of the Pareto front. On the other hand, QNSGA-II maintained high degrees of the diversity of their solutions during the search of the best optimal solution for either oil spill footprint detection or oil spill spreading level in SAR data. In this regard, the QNSGA-II is able to better distribute its population along the obtained front than Entropy and Q-MEA.

Moreover, QNSGA-II improves the quality of the non-dominated set accompanied by the diversity of the population in multi-objective problems of polarimetric SAR features. In fact, QNSGA-II is proposed by employing the concept and principles of quantum computing, for instance, uncertainty, superposition and interference. Indeed, QNSGA-II maintains a better spread of non-dominated set and accurately locate the solutions of automatic detection of oil spill spreading which are close to the Pareto-optimal front (Fig. 13.21). In other words, QNSGA-II improves proximity to the Pareto-optimal front, preserving diversity integral by retaining compensations of Quantum-inspired Evolutionary Aalgorithm (QEA). In this understanding, improving proximity revenues to locate the superior solutions which are appraised as virtuous individuals by a fitness function.

13.13 Comparison with Previous Studies

This study differs from the previous work performed by Marghany and van Genderen [261] and because the recent work presents an automatic classification based on QNSGA-II, whereas the study performed by Marghany and van Genderen [261] used an approach that is considered to be a semi-automatic tool for oil spill detection. In contrast to the previous study of Fiscella et al. [177], the Mahalanobis classifier provides an oil spill classification pattern in which slight oil spill pixels can be distinguished from medium and heavy oil spill pixels. Nevertheless, the findings of this study are consistent with the results of Topouzelis et al. [173]. The QNSGA-II algorithm was able to automatically extract oil spill pixels from the surrounding pixels without using a separate segmentation algorithm, as was done by Skrunes et al. [262]. Further, all the algorithms have been introduced are effectively depended on wind speed conditions. Nonetheless, the NSGA-II can automatically discriminate oil spill from the surrounding pixels even under wind speed of 6 ms^{-1}. Further, TCNNA algorithm of Garcia et al., [228] is based on entropy which first introduced by Marghany [163] as an excellent tool for oil spill detection in SAR data. Indeed, the capability of a SAR satellite to differentiate between oil, low wind areas, look-alikes is restrained by the noise floor of SAR. However, QNGSA-II explicit diversity preservation mechanism is involved in QNSGA-II is able to overcome this issue. The Support Vector Machine (SVM) was implemented by Matkan et al. [263] for automatic detection of the oil spill is based on thresholding and is not able to provide any information regarding the level of oil spill footprint spatial variation of thickness to lightness levels as compared to QNGSA-II. This work performs better discrimination of oil spill spreading as compared to fractal studies have been done by Marghany et al. [198,206]. In addition, this work proves the work done by Marghany [264] which claimed that the NSGA-II algorithm can be used to investigate the level of oil spill spreading in single SAR satellite data. QNGSA-II delivers an excellent promise to monitor oil spill spreading from full Polarimetric SAR images such as UAVSAR (uninhabited aerial vehicle synthetic aperture radar) and RADARSAT-2 quad-polarization and TerraSAR-X.

Simulation of Trajectory Movements of Oil Spill in Multisar Satellite Data Using Quantum Hopfield Algorithm

The previous chapters delivered a comprehensive intellect of the SAR mechanism for imaging and automatic detection of an oil spill from the point of view of quantum mechanics and quantum machine learning. This chapter critically evaluates the potential of Synthetic Aperture Radar's (SAR) and quantum machine learning for modeling the trajectory movement of oil spills. Unquestionably, SAR techniques are intensively used for monitoring and modeling the sea surface physical properties, for instance, wave spectra, current movements and bathymetry mapping. The main question which can, of course, be raised is how sequences of SAR images can simulate the trajectory movement of an oil spill without involving nonlinearity of the Doppler frequency model?

14.1 Oil Spill Trajectory Models: What They are, How They are Created, How They are Used an Oil Spill?

The trajectory model is a computer creating a prediction of the conceivable fate and behavior of an oil slick on the surface of the ocean. Information on ocean currents, wind speed and direction, and the oil sort are the main data input to the software that generates the model. These models are used for preparedness training in exercises that drill a specific response scenario. In an actual oil spill response trajectory models are generated early on to help guide response operations. The models are then updated frequently using visual and remote sensing observations of the oil slick, wind and ocean conditions. The oil spill trajectory model delivers prediction of how an oil spill spreads under the impact of wind, currents and other coastal hydrodynamic components. It also predicts the uncertainty of oil spill trajectory movements. Moreover, it monitors the change of chemical and physical properties of an oil spill during the time that it remains on the water surface. Therefore, the precise oil spill models are based on understanding the mechanisms that disturb oil fate and trajectory. In fact, these mechanisms are extremely complex, which are governed by transport processes and oil weathering processes. Oil transport, therefore, is achieved through spreading dispersion and sedimentation and oil weathering occur owing to evaporation, emulsification, dissolution, biodegradation and photo-oxidation [268]. Nonetheless, oil spill trajectory models involve uncertainties which are seen in the prediction motion, and the tendency for the oil to break into smaller slicks and information about initial and discharge conditions, which are excluded in the most available oil spill modeling software. Environmental factors, therefore, also play a foremost role in the fate of an oil slick, by affecting the oil weathering processes, their occurrence or significance [269].

Conventional oil spill trajectory models are deterministic and deliver a "best guess" of oil movement and fate; nevertheless, as scientific acquaintance develops, which it has become evident that some indication of the model uncertainty is necessary. In this regard, uncertainty arises from the input and environmental data necessary for the model operation, but also relies on the length and time scale of the spill [269]. However, if some of the details of the input data are missing, the use of such a sophisticated model will invariably lead to results that will deviate from the correct ones by an uncertain value [273].

The ambiguity in the trajectory is a consequence of the uncertainty in the forecasting of the forthcoming meteorological data and is liberated of the sort of oil spill model. Currently, weather predictions have virtual analytical skill for a period up to 4 days. This, however, reduces steadily as time grows [270–271].

Presently, there are numerous oil spill models with various degrees of complexity to compute the trajectory of spills. Nonetheless, the accomplishment of the use of every model is reliant on the mathematical formulation of the model itself, however, and on the precision of the input data, and lastly on how the outcomes are clarified. In this view, the input data are conditional on countless causes of error and approximately of the environmental parameters have stochastic performance. Henceforth, there is roughly the same amount of vagueness combined with the input variables that should not be abandoned [270,273].

The foremost fault of oil spill models originates from the application of the mathematical formulation of spreading owing to Fay [274]. In fact, Fay's algorithm was developed basically for calm sea circumstances and it contains formulas to regulate the growth of a circular spill, considering the physical-chemical properties of the crude. Even though Fay's algorithm can be applied to compute the area impartial after the spill. In actual fact, crude oil correspondingly spreads through the stroke of the currents and the turbulence that exists in the upper layers of the sea. Consequently, an excellent approach would be to compute the spreading in view of the laws of turbulence of the ocean. Nonetheless, in this case , additional models and input data are required, which can develop more uncertainties [273].

14.2   Role of Synthetic Aperture Radar for Tracking Oil Spill Trajectory Movement

Synthetic Aperture Radar (SAR) and spill tracking devices are, nonetheless, no substitute for trajectory analysis. While SAR is useful in determining spill location, it is a tool that only indicates the oil slick's present location and not where it is going in the future. To properly deploy cleanup resources, future spill locations, and possible shoreline impacts need to be anticipated. Also, all approaches of observing and tracking the oil have limitations on when they can perform properly. If environmental or spill conditions are outside the sensor's window of operation, trajectory analysis provides an estimation of the current position of the oil slick in the interim. Sensors, such as SAR, can provide a potential tool for forecasting the oil slick trajectory movements. SAR can provide information on the geophysical parameters which impact on the oil slick. The question can arise as to how the oil slick trajectory movement can be predicted by utilizing sequences of multiSAR sensors?

Zheng et al. [266] demonstrated that space shuttle photography can be used for analyzing oil pollution in the open ocean. Zheng et al. [266] determined the time of an oil slick being discharged into the sea by dividing the distance from the location of the oil slick to the ship by the shipping speed. By using this method, Zheng et al. [266] were able to obtain the dependence of the oil slick width at the time, which characterizes the oil slick spread. Finally, Zheng et al. [266] determined the oil slick trajectory movements by using the oil slick width information. Zheng et al. [266] found that the oil slick width increased with time. However, its utility for monitoring oil spill trajectories is limited by the satellite revisit time and the swath of the SAR. Furthermore, it is a challenge to distinguish oil spills from other look-alike natural phenomena (biogenic slicks, upwelling, low wind areas, etc.) in SAR images. Hence, it is desirable to combine SAR images with model simulations to gain more accurate oil spill information [267]. Recently, personalized oil spill information which is obtained from multiSAR data using the GNOME model to predict oil spill trajectories [275]. Yet, there is no study that implements the full capability of SAR images to predict the oil spill trajectory movements as a function of wind and current speed, which can be retrieved from SAR images.

Excellent identification of the oil slick trajectory is significant for early warning and contingency planning. The classical method for oil slick trajectory model cannot be ignored. These classical methods can be used as the basic tool for improvements to the SAR oil slick trajectory model. The most widely used spreading formula is Fay's algorithm. In fact, Fay's algorithm, involves two expressions, which are used for the spreading of the thick oil, and the total area of thick and thin slicks. For the thick slick, spreading consists of two parts: first a loss of the area due to oil flowing from thick to thin slicks and the second corresponding to the gravity-viscous phase of spreading. Fay's algorithm detects the change of the area of the thick slick per unit time. These spreading mechanisms are a function of the change of surface current movements.

14.3 Hypotheses and Objectives

This chapter focuses on the following approaches: (i) that **quantum Hopfield** can be used to discriminate the oil slick area from the surrounding waters. **Quantum Hopfield** could be used to map the location of oil slick on the coastal water, and (ii) that the integration between **quantum Hopfield** could be used as an automatic tool for oil spill detection; (iii) that **quantum Hopfield can** deliver a real-time current movement in the sequences of multiSAR data; (iv) that Loop Current is the main parameter affecting oil slick trajectory in the sequences of SAR images; (v) how far will the implementation of quantum Hopfield algorithm to multiSAR data to track the oil slick trajectory's path?

The main objective of this chapter is to operate quantum Hopfield for automatic detection of the oil slick and detect the oil slick trajectory model. This can be divided into four objectives: (i) to utilize quantum Hopfield for automatic detection of oil slick; (ii) to model the physical properties of surface current from multiSAR images; (iii) to model the effect of surface current on the oil slick pattern; (iv) to predict the oil slick trajectory model from multiSAR data based on quantum Hopfield algorithm.

14.4 Fay's Algorithm and Trajectory of Oil Spill

The oil spill simulation model that is exploited contains the modules to compute the weathering of the spill, its trajectory and a database where the required information is encoded as qubits. The database has geo-referenced information of the bathymetry and the characteristics of the most common crude oils required from multiSAR data. At the moment the data covers the Deepwater Horizon (DWH) oil incident, one of the world's largest incidental oil pollution event that occurred in the Gulf of Mexico, with a resolution of 1×1. The geophysical parameters of wind speed and ocean current surface are retrieved from the multiSAR data. In so doing, the quantum neural Hopfield algorithm is implemented to simulate ocean current condition based on the oil slick pattern morphology for each SAR image.

Consistent with Fay's algorithm, the spreading of oil spill involves three phases [274]. Instantaneously after a spill, the oil slick is reasonably thick [276]. Consequently, in the initial phase, gravity and inertial forces direct the spreading process, with gravity being the accelerating force and inertia the retarding force. In the second phase, gravity and viscous forces rule the spreading, with the viscosity being the retarding force. A third phase is eventually reached in which interracial tension and viscous forces impose the spreading [277]. In this view, Fay's algorithm is implemented to simulate the oil slick parcel movements. In this regard, Fay's algorithm is a numerical model which is used to describe the oil slick spreading. Based on Fay's algorithm, the spreading process, which is controlled by inertia, viscosity and surface tension. As stated by Fay's algorithm, the oil slick is assumed to spread as an ellipse with the major axis in the azimuth direction and minor axis in the range direction of SAR data (Fig. 14.1). Consistent with Marghany [278] the area of the ellipse can be given by:

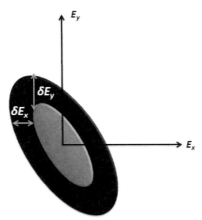

Figure 14.1. Elliptical spreading of the oil spill.

$$A = \left(\frac{\pi}{4}\right) E_x E_y \tag{14.1}$$

here $E_{x,y}$ is the major and minor axis in the azimuth and range directions, respectively. In order to estimate the major and minor value of the axis, Marghany [278] developed the oil spill trajectory algorithm for SAR data, which is given by:

$$\Delta E_x = \Delta E_y + CV^{1.3} \left[\frac{2v_{sx}v_{an}^2}{\Delta f_i \lambda R_T}\right] 0.75 \left[\frac{f_i \lambda R_T}{2v_{an}^2}\right]^{0.25} \tag{14.2}$$

Marghany [278] demonstrated that the major axis of oil slick spreading (Equation 14.2) can be based on the time variation along the azimuth direction. This time variation is a function of the Doppler frequency shift f_i and the velocity of the oil slick along the azimuth direction (v_{sx}), and current velocity v_{cx} in the azimuth direction, which can be retrieved from SAR data. The displacement along the range direction can be derived from Equation 14.2, then ΔE_r can be given by

$$\Delta E_r = C_2 v_{cy}^{1.3} \left[\frac{v_{cy} \sin \gamma R_T}{v_{an}^2}\right] \tag{14.3}$$

The forecasting trajectory model is defined as the model used to determine the slick parcel movement over time due to the effect of surface current movement [278]. The spreading displacement (S) of an oil particle in the azimuth and range direction (x, y), respectively during a time dt can be given by

$$dx = S_x dt \tag{14.4}$$

$$dy = S_y dt \tag{14.5}$$

$$\text{where } S = \frac{E_y^3 E_x' x^2 + E_x^3 E_y' y^2}{E_x E_y (E_x^2 y^2 + E_y x^2)} \tag{14.6}$$

$$E_x' = dE_x / dt \tag{14.6.1}$$

$$E_y' = dE_y / dt \tag{14.6.2}$$

Once the current velocities and the rate of spreading are modeled from SAR data, it is easy to establish the trajectory model of the oil slick. The oil slick trajectory movements can be derived from Equation 14.6 by using the Lagrangian model (Fig. 14.2). The parcels of oil slicks are divided into a large number of Lagrangian parcels of equal size. At each time step, each parcel is given a diffusive and a convective displacement as follows. We assumed that the initial position (x_i, y_i) of spreading $S(t)_i$ in the position of the 1th parcel and the first guess position of oil slick spreading is S' $[(t + \Delta t)_i]_{i+1}$. The velocity vectors are linearly interpolated in both space and time. The first guess position can be given by

$$S'(t + \Delta t)_{i+1} = S(t)_i + [V(S(t + \Delta t)_i]_{i+1,j} \tag{14.7}$$

and the final position can be given by

$$S(t + \Delta t)_{i+1,j} = S(t)_{i,j+1} + 0.5[V(S,t) + V(S',t + \Delta t)]_{i+1,j} \Delta t \tag{14.8}$$

Trajectories are terminated if they exit the model down to the water depth, but the advection continues along the surface. The integration time step Δt can vary during the simulation. Using the Lagrangian interpolation from the four surrounding grid nodes (Fig. 14.2) the current velocity components in the azimuth direction and range direction can be derived. This is modulated by sea truth data. The current velocity in the azimuth and in the range directions are V_{cx} and U_{cy}, respectively, and can be given by:

$$\left(\frac{\partial^2 V_{cx}}{\partial^2 x}\right) = \frac{v_{i+1} - 2v_{i,j} + v_{i-1}}{L} \tag{14.9}$$

$$\left(\frac{\partial^2 U_{cy}}{\partial^2 y}\right) = \frac{u_{i+1} - 2u_{i,j} + u_{i-1}}{L} \tag{14.10}$$

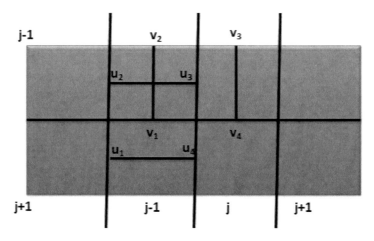

Figure 14.2. A grid of retrieving surface current using the Lagrangian model.

The conventional trajectory of the spill is mathematically expressed by:

$$V_R = V_c + w_f \times V_W \qquad (14.11)$$

The trajectory of the oil spill is a function of the resultant drift velocity, V_R, which is the vectorial addendum of the local current, V_c, to a fraction, w_f, of the wind velocity V_W. The wind velocity has approximately 4% effect on the trajectory of the oil spill. The velocity of the current, V_c, is supposed to include all components of the local current, except the wind-induced surface component. Therefore, excluding the last one, V_c will be the resultant of the vectorial addition of the large-scale mean circulation (residual current) and the tidal currents [273]. Therefore, a drift angle can also be deliberated either reliant on the wind velocity or constant, which ranges between 0° to 20°. In more specific terms, the drift angle is predictably ranged between 10 to 17° [278]. In this regard, the drift angle can be estimated by:

$$\theta = 25e^{\left(\frac{-10^{-8} V_W^3}{gv}\right)} \qquad (14.12)$$

Equation 14.12 is developed based on field observations. It is interesting to note e that the gravitational acceleration g and the kinematic viscosity of seawater v direct the drift angle. Conversely, the lack of information about the precise deflection angle and the ambiguity of the wind factor of 2–4% represent other reasons to consider the model uncertainty.

14.5 Hopfield Neural Network for Retrieving Current Pattern from MultiSAR Data

Neural network techniques are an artificial intelligence procedure that belong to the expert system and knowledge-based approaches to learning [279,280]. On the contrary, scientists agreed that neural networks belong to the same classes of techniques as automated pattern recognition, stereo vision, motion analysis and object tracking problems [281,282]. In this context, scientists have defined a Hopfield model as a kind of neural network. The Hopfield network has no special input or output neurons, but all are both input and output, and all are connected to all others in both directions (with equal weights in the two directions). Input is applied simultaneously to all neurons, which then output to each other and the process continues until a stable state is reached, which represents the network output [283].

Consistent with Côté and Tatnall [282], Hopfield neural networks are considered as a promising method for determining a minimum of the energy function. For instance, motion analysis and object pattern recognition might be coded into an energy function. Furthermore, the actual physical constraint, heuristics, or prior knowledge of objects and the system can be coded into the energy function. A pattern, in the context of the N node Hopfield neural network is an N-dimensional vector $V = (v_1, v_2, \ldots v_n)$ and $U = (u_1, u_2, \ldots u_n)$ from space $S_p = \{-1,1\}^N$. A special subset of S_p is set of exemplar $E = \{e^k : 1 \le k \le K\}$, where $e^k = (e^k_1, e^k_2, \ldots, e^k_n)$ and k is exemplar pattern where $1 \le k \le K$. The Hopfield net associates a vector from S_p with an exemplar pattern in E.

In this context, the net partitions S into classes whose members are in the same way of exemplar pattern that presents the classes. For the sea surface current features in the multiSAR satellite images (f), and $f_t(i) \in \{-1,1\}$ to represent neuron states, i.e., either -1 or $+1$, which serve as processing units. Each neuron has a value at time t denotes by $f_t(i)$. The permanent memory of neural set resided within the interconnections between its neurons which is named by the strength of the synapse (w_{ij}) or connection between two pair of the neuron $f(i)$ and $f(j)$, According to Nasrabadi and Choo [283], the design specifications for this version of Hopfield net require that $w_{ij} = w_{ji}$ and $w_{ii} = 0$. Following, Yi et al. [14], the propagation rule τ_i, which defines how neuron sates and weight combined as input to a neuron can be described by:

$$\tau_i = \sum_{j=1}^{N} f_t(j) w_{ij} \tag{14.13}$$

The Hopfield algorithm consists of (i) assigning weights of synaptic connections, (ii) initializing the net with unknown pattern, (iii) iterating until convergence and continuing features tracking (Fig. 14.3). The first step of assigning a weight w_{ij} to synaptic connection can be achieved as follows:

$$w_{ij} = \begin{cases} \sum_{k=1}^{K} e_i^k e_j^k & \quad i \neq j \\[2mm] & \text{if} \\[2mm] 0 & \quad i = j \end{cases} \tag{14.14}$$

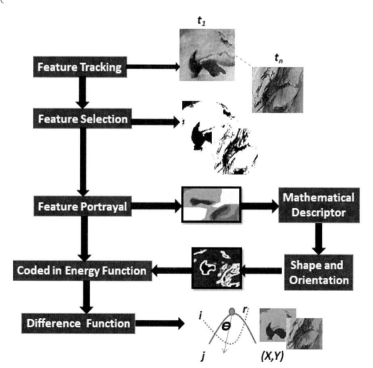

Figure 14.3. Block diagram of feature tracking by Hopfield algorithm of step 1.

In the second step, the pattern that is to be quantified is introduced to the net. If the vectors $V = (v_1, v_2, \ldots v_n)$ and $U = (u_1, u_2, \ldots u_n)$ are unknown patterns, then set

$$f_0(i) = v_i, \ 1 \le i \le N \tag{14.15}$$

$$f_0(j) = u_j, \ 1 \le j \le N \tag{14.16}$$

The third step involves the estimation of next state values for the neuron in the set using the propagation rule and activation function F. This step can be expressed mathematically by the given formula:

$$f_{t+1}(i) = F\left(\sum_{j=1}^{N} f_t(j) w_{ij}, f_t(i)\right) \tag{14.17}$$

Repetition of second and third steps is involved as the fourth step to perform quantification of another sea surface current pattern in multiSAR images (Fig. 14.4).

Hopfield neural network could be identified as current pattern featured by mathematically comparing each other in order to build an energy function. Consistent with Côté and Tatnall [282] the difference function to determine the discriminations between different features f_i, f_j by a given formula:

$$diff(f_i, f_j) = G.\max\left|\max(\frac{l_i}{l_j}, \frac{l_j}{l_i}) - L'', 0\right| + H.\max[\min(\left|\theta_i - \theta_j\right|, 2\pi - \left|\theta_i - \theta_j\right| - \theta'', 0]$$

$$+ J.\max\left|dis_j - dist'', \theta\right| \tag{14.18}$$

where L'' is curvature shape of the current feature, dis_{ij} is the distance between sea surface current features f_i and f_j, and G and H and J are constants and is an angle of orientation of local curve element θ. In addition, $dist''$ and θ'' are the minimum acceptable distance and the maximum acceptable rotation angle, respectively before energy function. In practice, the Hopfield neural network can be quantified by the Euler equation [282] of motion which can be expressed as

$$\frac{dW_i}{dt} = \sum_{j}^{N} \tau_{ij}.(F\left(\sum_{j=1}^{N} f_t(j) w_{ij}, f_t(i)\right) \lambda_i W_i) + B = \frac{\partial E}{\partial f_{t+1}(i)} \tag{14.19}$$

where B is the external bias on the neuron i, λ is the steepness of the function and E is the network energy, Wang et al. [285]; Juang [286]; Arik [287] which can be defined as:

$$E(V, U) = -0.5(v_1, v_2 \ldots, v_n) \begin{pmatrix} w_{11} & w_{12} & \ldots & w_{1k} \\ w_{21} & w_{22} & \ldots & w_{2k} \\ . & & & \\ . & & & \\ w_{n1} & w_{n2} & \ldots & w_{nk} \end{pmatrix} \begin{pmatrix} u_1 \\ u_1 \\ : \\ : \\ u_k \end{pmatrix} \tag{14.20}$$

Following Côté and Tatnall [282], the minimization of energy Equation 14.20 can be used for sea surface current features tracking and stereo matching. The contribution of this chapter is to determine the current velocity from sequences of multiSAR images to simulate the trajectory of the oil spill model. Sequences of point-like current features are selected as candidates for sea surface current movements in the multiSAR images (Fig. 14.3). The adjoining pixels of current features have to fill the following criteria: (i) the mean energy of the current pixels must be higher than the mean energy of the surrounding pixels, and (ii) the number of sea current pixels should lie between a lower and upper bound. Finally, Hopfield algorithm is implemented on the detected oil spill and current features, a displacement vectors are estimated by determining the center position of sequences of frames generated over candidate current feature (Fig. 14.5). In Fig. 14.5, the gray circles represent the neuron output, where the brighter the circle, the higher the output [288–290].

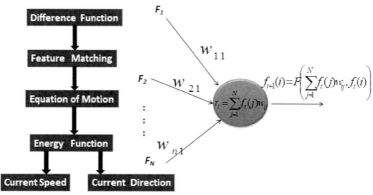

Figure 14.4. Block diagram of feature tracking by the Hopfield algorithm of step 2 and Hopfield neural network is used in the study.

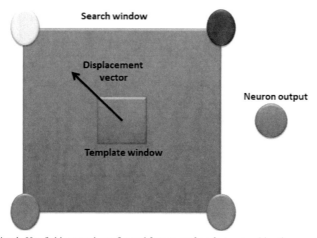

Figure 14.5. Simple Hopfield network configured for sea-surface feature tracking from multiSAR images.

14.6 Quantum Hopfield Algorithm

Let us consider the mission of expending multi-qubit quantum systems to create quantum neural networks. In this regard, a direct association between neurons and qubits [291], unlocking access to quantum properties of entanglement and coherence of multiSAR data is implemented. In this view, let us instead encode the neural network into the amplitudes of a quantum state. This is achieved by hosting an association regulation between activation patterns of the neural network and pure states $|f\rangle$ of d-level of a quantum system. In this understanding, the determined state features of the oil spill and current pattern can be expressed as $f \rightarrow |f\rangle |f|_2$ with $|f|_2 = \sqrt{\sum_{i=1}^{d} f_i^2}$ and $|f\rangle := |f|_2^{-1} \sum_{i=1}^{d} f_i |i\rangle$, which is written with respect to the standard basis such that $\langle f|f\rangle = 1$. The d-level quantum system can be implemented by a register of $N = [log_2 d]$ qubits so that the qubit overhead of representing such a network scales logarithmically with the number of neurons [291].

The quantum Hebbian learning algorithm (qHeb) is used for weighting matrix W of the Hopfield network. It relies on two significant perceptions: (i) a direct association of the weighting matrix with a mixed state ρ, and (ii) one can efficiently perform quantum algorithms that harness the information contained in W. Therefore, ρ registers of N qubits in relation to:

$$\rho := \omega + \frac{\prod_d}{d} = M^{-1} \sum_{m=1}^{M} |f^{(m)}\rangle \langle f^{(m)}| \qquad (14.21)$$

where $\prod_d$ is the d-dimensional identity matrix and M is a training set of activation patterns f, with $m \in \{1, 2, ..., M\}$. In this view, the Hopfield neural network can be taught training set using the Hebbian learning rule, which sets the weighting matrix elements ω_{ij} along with the number of occasions in the training set that the neurons i and j mix together. In this sense, Hebbian learning is employed to set the weighting matrix W from d-length training data $\{F^{(m)}\}_{m=1}^{M}$. In this regard, the qHop algorithm continues to compute $|v\rangle = \frac{1}{A_{SAR}} |w\rangle$ where the pure state $|w\rangle$ is primarily arranged, which comprises user-identified neuron thresholds of the SAR data inverse matrix $\frac{1}{A_{SAR}}$ of feature identifications and a partial memory pattern. In other words, the inverse matrix $\frac{1}{A_{SAR}}$ containing information on the training data and regularization γ. To this end, the output of the pure state $|v\rangle$ contains information on the reconstructed state of SAR features $|f\rangle$ and Lagrange multiplier vector λ. The feature state matrix of SAR data can be expressed as a quantum state by:

$$A_{SAR} |v|_2 |v\rangle = |w|_2 |w\rangle \qquad (14.22)$$

where

$$|v\rangle := \frac{1}{|v|_2} \left(|f|_2 |0\rangle \otimes |f\rangle + |\lambda|_2 |1\rangle \otimes |\lambda\rangle \right) \qquad (14.22.1)$$

$$|w\rangle := \frac{1}{|w|_2} \left(\left|\Xi\right|_2 |0\rangle \otimes |\Xi\rangle + \left|f^{(inc)}\right|_2 |1\rangle \otimes \left|f^{(inc)}\right\rangle \right) \tag{14.22.2}$$

where $f^{(inc)}$ is the normalized quantum state corresponding to the incomplete activation pattern and $\left|f^{(inc)}\right|_2 = l$. The objective of the Equations 14.22.1 and 14.22.2 is to optimize the energy function E in Equation 14.20. In addition, Ξ presents each element of qHop which should be set so that its magnitude is of order at most 1 qubit. In other words, $\Xi =: \{\Xi_i\}_{i=1}^{d} \in \mathbb{R}^d$ is a user-specified neuronal threshold vector that determines the switching threshold for each neuron. Let us identify the set of M unitary operators $\{U_k\}_{k=1}^{M}$ acting on an $N + 1$ register of qubits consistent with

$$U_k := |0\rangle\langle 0| \otimes \Pi + |1\rangle\langle 1| \otimes e^{-i\left|f^{(k)}\right\rangle\left\langle f^{(k)}\right|\Delta t} \tag{14.23}$$

Equation 14.23 reveals that the unitaries implement various pattern detections in multiSAR data for a small difference time Δt under the circumstance of $\left|f^{(k)}\right\rangle\left\langle f^{(k)}\right|$. Conversely, these unitaries can be simulated using the following mathematical equation:

$$U_s := e^{-i|1\rangle\langle 1|\otimes s\Delta t}$$
$$|0\rangle\langle 0| \otimes \Pi + |1\rangle\langle 1| \otimes e^{-is\Delta t} \tag{14.24}$$

here, $|1\rangle\langle 1| \otimes s\Delta t$ is 1-sparse and efficiently simulatable. Conversely, the trace tr_2 of U_s is over the second subsystem containing the state feature $|f\rangle$ in the multiSAR images, which is expressed by:

$$tr_2 \left\{ U_s \left(|q\rangle\langle q| \otimes \left|f^{(k)}\right\rangle\left\langle f^{(k)}\right| \otimes \sigma \right) U_s^\dagger \right\}$$
$$= U_k \left(|q\rangle\langle q| \otimes \sigma \right) U_k^\dagger + \mathcal{O}\left(\Delta t^2\right). \tag{14.25}$$

In Equation 14.25, the subsystem of ancilla qubit $|q\rangle\langle q|$ and σ efficiently experiences time evolution $\mathcal{O}\left(\Delta t^2\right)$ with U_k. The unitary $e^{iA_{SAR}t}$ is simulated to fix the error ϵ for arbitrary t. Indeed ϵ can be estimated by:

$$\epsilon := \left\| \left(e^{-i\left|f^{(1)}\right\rangle\left\langle f^{(1)}\right|\frac{t}{nM}} ..e^{-i\left|f^{(M)}\right\rangle\left\langle f^{(M)}\right|\frac{t}{nM}} \right)^n - e^{-iA_{SAR}t} \right\| \in \mathcal{O}\left(\frac{t^2}{n}\right). \tag{14.26}$$

Equation 14.26 indicates that the repetition of $n \in \mathcal{O}\left(\frac{t^2}{\epsilon}\right)$ is required alongside with M-sparse Hamiltonian simulations. The advantages of this approach are that the copies of the training states $\left|f^{(m)}\right\rangle$ can be used as quantum software states and, in addition, it does not require superpositions of the training states. Conversely, the multiSAR feature matrix A_{SAR} can be identified as:

$$A_{SAR} := \begin{pmatrix} 0 & P \\ P & 0 \end{pmatrix} + \begin{pmatrix} -\left(\gamma + d^{-1}\right)\Pi_d & 0 \\ 0 & 0 \end{pmatrix} + \begin{pmatrix} \rho & 0 \\ 0 & 0 \end{pmatrix} \tag{14.27}$$

In this regard, the simulation time is split into n small time steps $t = n\Delta t$. In this case, the error ϵ can be extended into Taylor expansion as:

$$\epsilon_{\Delta t} := \left\| e^{iA_{SAR}\Delta t} - U_B\left(\Delta t\right)U_C\left(\Delta t\right)U_D\left(\Delta t\right) \right\| \in \mathcal{O}\left(\Delta t^2\right) \tag{14.28}$$

where $U_B\left(\Delta t\right), U_C\left(\Delta t\right)$, and $U_D\left(\Delta t\right)$ are the operators of $\begin{pmatrix} 0 & P \\ P & 0 \end{pmatrix}$, $\begin{pmatrix} -\left(\gamma+d^{-1}\right)\Pi_d & 0 \\ 0 & 0 \end{pmatrix}$, and $\begin{pmatrix} \rho & 0 \\ 0 & 0 \end{pmatrix}$, respectively. In this understanding, at most $\mathcal{O}\left(\Delta t^2\right)$, the errors $e^{i\begin{pmatrix} 0 & P \\ P & 0 \end{pmatrix}\Delta t}$, $e^{i\begin{pmatrix} -\left(\gamma+d^{-1}\right)\Pi_d & 0 \\ 0 & 0 \end{pmatrix}\Delta t}$, and $e^{i\begin{pmatrix} \rho & 0 \\ 0 & 0 \end{pmatrix}\Delta t}$ are simulated respectively, which allow simulating the unitary $e^{iA_{SAR}t}$ to an error of $\epsilon \in \mathcal{O}\left(n\Delta t^2\right)$. Then the matrix of spectral energy of the qHop can be mathematically expressed as:

$$A_E = \sum_{j:|\lambda_j(A_{SAR})|>\lambda} \lambda_j(A_{SAR})\left|E_j(A_{SAR})\right\rangle\left\langle E_j(A_{SAR})\right| +$$
$$\sum_{j:|\lambda_j(A_{SAR})|<\lambda} \lambda_j(A_{SAR})\left|E_j(A_{SAR})\right\rangle\left\langle E_j(A_{SAR})\right| \tag{14.29}$$

Equation 17.29 demonstrates how to simulate the gradient energy of the various feature pattern variations in the multiSAR as a function of the size of the eigenvalues λ_j in comparison to a fixed user-defined number $\lambda > 0$. In this circumstance, qHop maintains the polylogarithmic efficiency in run time whenever λ is such that $\lambda^{-1} \in O(poly(\log d))$. Therefore, the primary matrix inversion algorithm returns (up to normalization) as:

$$A_E^{-1}|E\rangle = \sum_{j:|\lambda_j(A_{SAR})|\geq\lambda} \left\langle E_j(A_{SAR})\middle|E\right\rangle\left(\lambda_j(A_{SAR})\right)^{-1}\left|E_j(A_{SAR})\right\rangle\left(\lambda_j(A_{SAR})\right)^{\cdot} \tag{14.30}$$

The input state of the energy gradient $|E\rangle$ is first achieved and it contains the threshold data and incomplete activation pattern) and considers the eigenbasis of the matrix of multiSAR image energies A_E. Therefore, obtaining information from each energy pattern require $O(K)$ operations of qHop. The qHeb algorithm is then initialized along with sparse Hamiltonian simulation [38] to perform quantum phase estimation, allowing $\sum_j \left\langle E_j(A_{SAR})\middle|E\right\rangle\left|\tilde{\lambda}_j(A_{SAR})\right\rangle \otimes \left|E_j(A_{SAR})\right\rangle$ to be obtained with $\tilde{\lambda}_j(A_{SAR})$ an approximation of the eigenvalue $\left(\lambda_j(A_{SAR})\right)$ to precision ϵ.

14.7 Quantum Trajectory Search for Oil Spill Movements in MultiSAR Data

The classical algorithms to solve the problem of oil spill modeling (Equations 14.8 and 14.11) take input in the form of a matrix to say S_R such that $\left[S_R\right]_{ij} = \phi_{ij}$, where ϕ_{ij} is the cost/distance/time or any other quantity taken to propagate from location i to location j in the multiSAR data. This quantity for the overall trajectory has to be minimized. Without loss of generality, the quantity as the cost is considered to obtain the trajectory movement of oil spill parcels. In this regard, the foremost motivation to consider the input as phases stems from the succeeding dualistic facts. Foremost, the matrix composed of the identified distances using the above procedure is not unitary in general, which entails that the performance and handling of the operator is not conceivable on a quantum computer. Subsequently, the phases will acquire paths

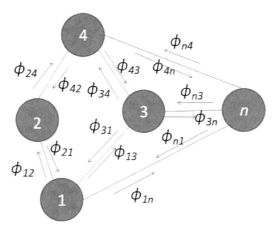

Figure 14.6. Oil spill trajectory movements over *n*-locations.

when tensor products of states are multiplied or considered with these phases as coefficients, that is the distances will get added as phases which are required for the search of accurate oil spill trajectory movements. Henceforth the input of the oil spill trajectory movement (Fig. 14.6) can be represented as a new matrix the input $\mathbb{Z} = e^{i\left(\phi_{ij}\right)}$.

Let us consider the quantum computation only as Oracle $O\left(\sqrt{N}\right)$, which is needed to solve the trajectory movement of the oil spill with probability 1. In this sense, it is required to find a solution ℓ of size N, which is known as the Oracle. Indeed $O\left(\sqrt{N}\right)$ is sufficient to solve the search problem of oil modeling by $O\left(\sqrt{\dfrac{N}{\ell}}\right)$. Hence the action of the Oracle O on the input $\mathbb{Z}$ can be signified by

$$O|\mathbb{Z}\rangle = (-1)^{f(\mathbb{Z})}|\mathbb{Z}\rangle \tag{14.31}$$

The action of the Hadamard transform $H^{\otimes q_n}$ on the input $|0\rangle^{\otimes q_n}$ is depleted to transform the input into a uniform superimposed state $|\psi\rangle$, is written as:

$$|\psi\rangle = H^{\otimes q_n}\left(|0\rangle^{\otimes q_n}\right) = \frac{1}{\sqrt{N}}\sum_{\mathbb{Z}=0}^{N-1}|\mathbb{Z}\rangle \tag{14.32}$$

Equation 14.32 reveals that all the q_n qubits are initially in the state $|0\rangle$. Then the Grover operator G is implemented for phase shift, which is transforming all non-zero basis state $|\mathbb{Z}\rangle$ to $-|\mathbb{Z}\rangle$ with $|0\rangle$ enduring unmovable. Therefore, the Grover operator G is basically the unitary transformation of the trajectory state as:

$$G = \left(H^{\otimes q_n}\left(2|0\rangle\langle 0|-I\right)H^{\otimes q_n}\right)O = \left(2|0\rangle\langle 0|-I\right)O. \tag{14.33}$$

The term $\left(2|0\rangle\langle 0|-I\right)$ is the unitary operator due to the phase shift. Equation 14.33, then can be enlightened as rotation in a two-dimensional space spanned by the vectors $|\alpha\rangle$ and $|\beta\rangle$, which is formulated as:

$$|\alpha\rangle = \frac{1}{\sqrt{N-M}}\sum_{\mathbb{Z},\mathbb{Z}\ \text{not marked}}|\mathbb{Z}\rangle \tag{14.34}$$

$$|\beta\rangle = \frac{1}{\sqrt{M}} \sum_{Z,Z \text{ not marked}} |Z\rangle \qquad\qquad (14.35)$$

In this sense, the initial state $|\psi\rangle$ for oil spill trajectory is in the span of $|\alpha\rangle$ and $|\beta\rangle$ as it is easily verified that $|\psi\rangle = \sqrt{\dfrac{N-M}{N}}|\alpha\rangle + \sqrt{\dfrac{M}{N}}|\beta\rangle$. Therefore, the number of iterations R is chosen such that $R\theta$ is the angle required to turn the state vector ψ very close to $|\beta\rangle$ and G is written as $\begin{pmatrix} \cos(\theta) & -\sin(\theta) \\ \sin(\theta) & \cos(\theta) \end{pmatrix}$ (Fig. 14.7). In this context, the

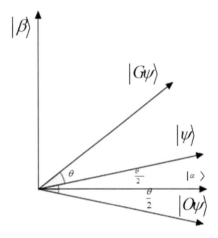

Figure 14.7. Grover operator as the rotation of the state vector $|\psi\rangle$.

algorithm proceeds by first selecting a random state from the input list and setting its associated cost as the running minimum [292–294].

The irregular m efficiently gear shifts the sum of Grover's rotation r that is achieved. A priori, it is not recognized that how numerous the states have their related costs less than the current running with minimum cost and therefore, m is adaptively mounted to guarantee that a precise value for r is designated as necessitated by Grover's search algorithm. In this view, Grover's search algorithm has been encoded in python language as follows:

```
# Bitstring Map as an algorithm input
SEARCHED_STRING = "1011010"
N = len(SEARCHED_STRING)
mapping = {}
for b in range(2 ** N):
    pad_str = np.binary_repr(b, N)
    if pad_str == SEARCHED_STRING:
        mapping[pad_str] = -1
    else:
        mapping[pad_str] = 1
# Connection
```

```
qc = get_qc('9q-qvm')

#=============================================================================
# Grove: Grove's Search Algorithm
#=============================================================================
# Run
algo = Grover()
ret_string = algo.find_bitstring(qc, bitstring_map=mapping)
print("The searched string is: {}".format(ret_string))
```

MSI laptop with CPU **of** Intel Core i9, GPU **of** Nvidia RTX 2070/2080, and **RAM of** 64GB is implemented to run Grover's search code.

14.8 Sequences of SAR Data for Oil Spill Trajectory Movements

In this chapter, observations from ENVISAT ASAR and COSMO-SkyMed (CSK) SAR are acquired. The European Space Agency (ESA) ENVISAT satellite was launched in March 2002, which operates in C-band with distinctive polarization combinations. Significant novel proficiencies of ASAR contain beam steering for acquiring images with diverse incidence angles, dual polarization and wide swath coverage. In this regard, observation swaths are ranged from 56 to 405 km in width, with resolutions from 30 to 150 m and at incidence angles from 15° to 45° [268]. The selected ENVISAT ASAR data were acquired from April 29th, 2010 to July 20th, 2010. Figure 14.8 refers

Table 14.1. Characteristics of ENVISAT ASAR and CSK SAR.

Sensors	Frequency	Polarization	Resolution	Swath Width	Revisit/ Repeat time
Envisat-ASAR (Not operational)	5.331 GHz (C-band)	HH/HV; VV/VH; VV/HH	4*20, 150 m	100, 400 km	2~3/35days
Cosmo-SkyMed Constellation	9.6 GHz (X-band)	HH, HV, VH, VV	1~100 m	10 km	1/16 days

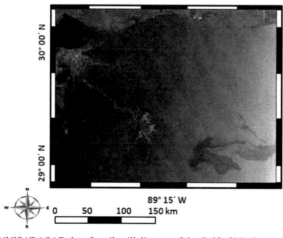

Figure 14.8. ENVISAT ASAR data for oil spill disaster of the Gulf of Mexico on April 29th, 2010.

to the Deepwater Horizon (DWH) incident that occurred in 2010 in the Gulf of Mexico, which captured by ENVISAT ASAR data on April 29th, 2010.

Therefore, one of the key disadvantages of SAR satellite imaging depends on its revisit time that, actual frequently, is not impenetrably adequate to guarantee a worthy example of the oil spill occurrence. In this regard, the COSMO-SkyMed (CSK) SAR provides excellent revisiting time. Within this context, the CSK constellation is attractive from an operational point of view, since it is a constellation of four X-band SARs, characterized by a very short revisit time, i.e., ≈12 hours, and it is able to

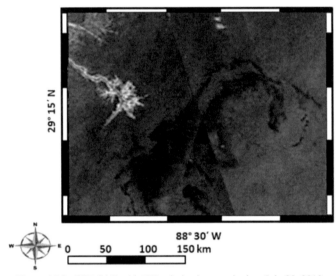

Figure 14.9. CSK SAR with HH polarization acquired on July 20, 2010.

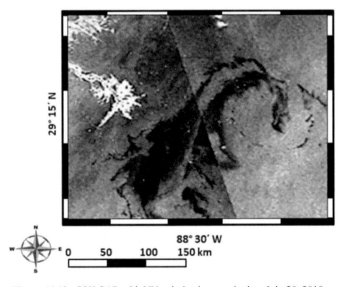

Figure 14.10. CSK SAR with VV polarization acquired on July 20, 2010.

operate in an incoherent dual-polarization mode (Ping Pong(PP) mode) [295,296]. The signal backscattered from slick-free and the oil-covered sea surface is analyzed using dual co-polarization, i.e., HH (Fig. 14.9)-VV (Fig. 14.10), ping pong mode under an incidence angle of 40° at mid-range with about a 15-m spatial resolution.

It is clear that the VV polarization provides excellent visualization for oil spill than HH polarization. This is demonstrated by both ENVISAT and CSK SAR data. In fact, VV polarization performs better at highlighting oil slicks from sea background for stronger power return, however, it may vary according to different oil types and sea conditions.

14.9 Trajectory Movements of Oil Spill Using Quantum Hopfield Algorithm

The quantum Hopfield algorithm delivers automatic detection of the oil spills and determines their spreading from a single image to sequences of SAR images over the different periods. In this view, quantum Hopfield demonstrates an early stage fluctuation of the oil spill spreading, which is captured by ENVISAT ASAR data from April 29th, 2010 (Fig. 14.11a) to May 9th, 2010 (Fig. 14.11c). It is worth mentioning that the pattern of oil spills are steady and predominantly seem as patches owing to widespread diffusion. Under the circumstances of diffusion, the oil spill changed its shape on May 2nd, 2010 (Fig. 14.11b).

In continuation with CSK SAR, the oil spill undergoes diffusion pattern development. Contrary to ENVISAT ASAR data, CSK SAR captured wide spreading of oil spill patches across the Gulf of Mexico from May 11th, 2010 (Fig. 14.12a) to May 14th, 2010 (Fig. 14.12c). CSK SAR can capture oil spill pattern development within 1-day revisit. In this view, Figs. 14.12b and 14.12c reveal wider spreading of oil spill than Fig. 14.12a. In this understanding, CSK reveals the wide development of oil spill pattern than early stage captured by ENVISAT ASAR data (Fig. 14.11).

Therefore, on July 20th, 2010, CSK SAR revealed thick patches of the oil slick without involving light or thin slick (Fig. 14.13). This could demonstrate the partial spreading of the oil spill as the light and thin oil spills washed ashore of Lousiana (Fig. 14.14).

Consistent with the above perspective, automatic detection of an oil spill from sequences of SAR done is successfully achieved using quantum Hopfield. In fact, the Hebbian learning rule taught a training set of the oil spill and other features of the Hopfield neural network. This assists the pure state $|w\rangle$ to be arranged in order to identify oil spill features as a function of identification neuron thresholds. In this regard, the pure state involves information on the reconstructed state of oil spill spreading in SAR data, which can be identified as a quantum state.

14.10 Impact of Current Pattern on Oil Spill Trajectory Movements

Consider a quantum system of the drifted oil spill particles moving from the point (V_i, t) to point (V_f, t) in the configuration space. In other words, the quantum Hopfield system at the initial moment of time in the state $|V_i\rangle = |v_i, t_i\rangle$, and at the final moment

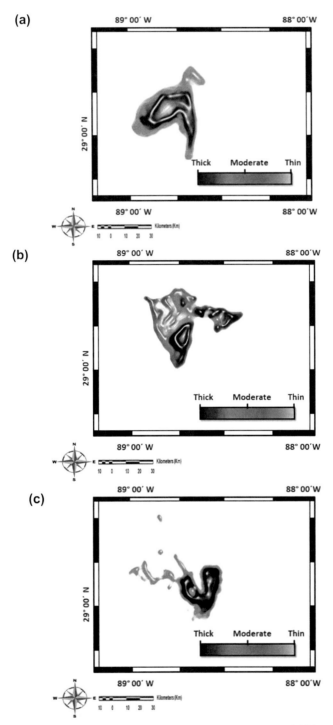

Figure 14.11. Automatic detection of oil spill spreading using quantum Hopfield from ENVISAT ASAR data (a) April 29th 2010, (b) May 2nd, 2010, and (c) May 9th, 2010.

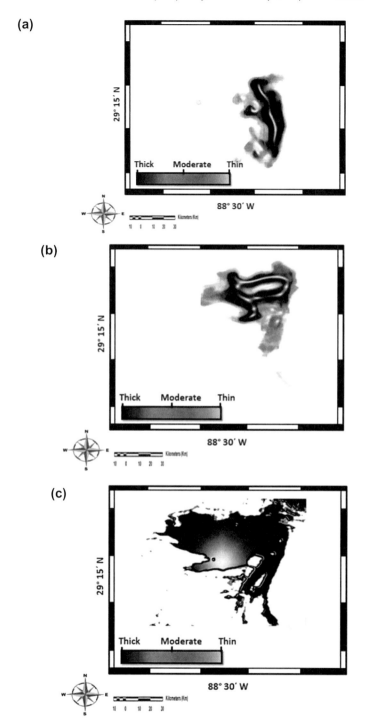

Figure 14.12. Automatic detection of oil spill spreading using quantum Hopfield from CSK SAR data (a) May 11th 2010, (b) May 13th, 2010, and (c) May 14th 2010.

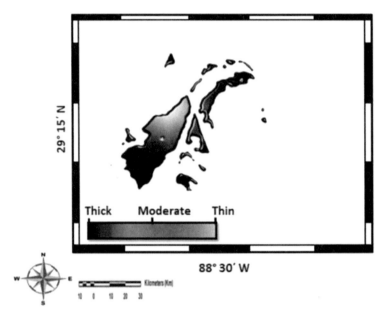

Figure 14.13. Patches of thick oil spill detected on July 20th, 2010 in CSK SAR.

Figure 14.14. Oil spill washed ashore of Lousiana.

in the state $|V_f\rangle - |v_f, t_f\rangle$. The scalar product of both states $\langle V_i | V_f \rangle$, which is also addressed as the transition amplitude. In this understanding, the transition amplitude can be presented as the path integral over all possible paths, which connect the initial configuration of a quantum Hopfield with its final configuration, which is given by:

$$\langle V_i | V_f \rangle = \int_{V_i(t)}^{V_f(t_f)} D\bar{\psi} D\psi \ e^{\left[\frac{i}{\hbar} \int_{t_i}^{t_f} \ell(v,\dot{v}) dt \right]}$$

(14.36)

This equation reveals Feynman's path integrals, which can define the fields of oil spill particles. In this view, $D\overline{\psi}D\psi$ is pointing at the integration over all possible configurations of oil spill trajectory path. Therefore, ℓ is the Lagrangian density. Due to the complexity of fermions, the Lagrangian density over both space and time can be expressed as an integral over the "hypervolume" by:

$$V_R\left[\overline{\psi},\psi\right] = \int_{t_i}^{t_f} dt \iiint d^3r\,\ell\left(\overline{\psi}\left(r,t\right),\psi\left(r,t\right)\right) \tag{14.37}$$

The quantum Hopfield was examined on ENVISAT ASAR and CSK SAR to simulate the trajectory movement of the oil spill based on the observed archived data, which are obtained from ROFFS™ (Roffer's Ocean Fishing Forecasting, Inc.) at http://www.roffs.com/deepwaterhorizon.html and Coastal Emergency Risks Assessment group (CERA, http://coastalemergency.org/). The wind and ocean current forcing is acquired from the operational products that are freely available globally. The quantum Hopfield can identify the pattern of the current flow along with the oil spill. Therefore, the mesoscale eddies are delivered clearly by using quantum Hopfield.

Indeed, the eddies are dominated features in the Gulf of Mexico. In this view, qHop simulates coherent local vectors across the oil spill. Figure 14.15 represents the predicted trajectory movement of an oil spill from April 29th, 2010 to May 10th, 2010. It is interesting to find that the qHop can forecast the oil spill spreading on May 10th, 2010 as absent of ENVISAT ASAR revisit.

Therefore, qHop reveals the great spreading of oil spill since May 14th, 2010 to May 27th, 2010 from CSK SAR (Fig. 14.16). The spreading was washed ashore under the impact of surface circulation. In fact, this circulation is influenced by cyclonic and anticyclonic eddies (ACs) as well as the Loop Current (LC) features that extend vertically to at least 800 m water depth [297]. However, the amount of surface oil

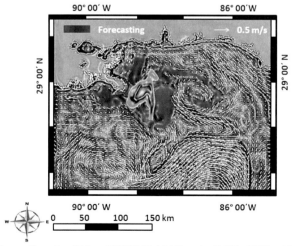

Figure 14.15. Forecasting oil spill from ENVISAT ASAR on April 29th, 2010 to May 10th, 2010 using the qHop algorithm.

Color version at the end of the book

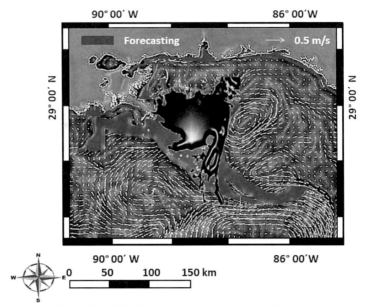

Figure 14.6. Oil spill trajectory movements over n-locations.

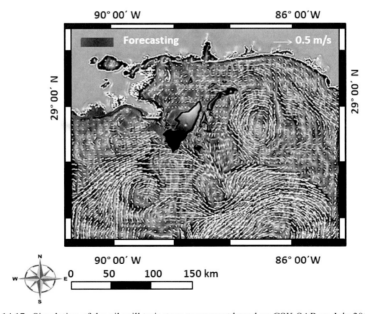

Figure 14.17. Simulation of the oil spill trajectory movement based on CSK SAR on July 20th, 2010.

Color version at the end of the book

entrained into the LC system appeared to be very small relative to the total amount of discharged oil, and the majority of the surface oil slicks were found around the well

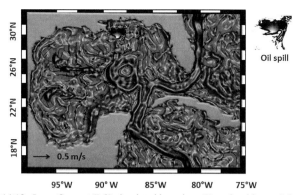

Figure 14.18. Loop Current (LC) simulated by using qHop algorithm on May 2010.

site and on the shallow shelf in May 2010 [298]. On the contrary, the oil spill tended to be arranged in small patches since July 20th, 2010 to July 29th 2010 (Fig. 14.17).

It is clear that the qHop identified accurately the edge of the loop currents, which involves several mesoscale eddies. In fact, the cycle of the Gulf Loop Current comprises four stages. The **first stage** of the cycle initiates with the Current shallowly moving the Gulf of Mexico from the Caribbean Sea, wrapping around Cuba and immediately heading out to the Atlantic. Over time, the Loop grows in size, reaching farther into the Gulf more and more in the **second phase** of the cycle. After reaching into the Gulf, a portion of the Loop detaches to create an eddy vortex that circles south of the U.S. Gulf Coast. This is the **third phase** of the cycle. This eddy starts to slowly drift toward Texas and Mexico, gradually decreasing in size and intensity. The Loop recedes back toward the western tip of Cuba during this stage. In the final **fourth phase,** the Loop retreats back to its original position and the eddy diminishes as it drifts west (Fig. 14.18).

14.11 How far the Impact of Loop Current on Oil Spill?

Some oil from the Gulf of Mexico spill was "gradually expected" to be towed into a robust current that was present on Florida's coasts. The sequences of the multiSAR data demonstrate that oil caught up in one of the eddies, and powerful whorls, which is attached to the Loop Current, a high-speed stream that pulses north into the Gulf of Mexico and travels in a clockwise pattern toward Florida's coasts. In this view, Fig. 14.16 shows that a "big, wide tongue" of oil extending south from the leading area of the spill, off the coast of Louisiana. Meanwhile, a particular eddy has intensified and expanded north in recent days. Figure 14.16 reveals that the eddy has ripped oil and dragged it southeastward approximately 100 miles (about 160 kilometers), which means the crude has circulated inward the Loop turbulent current. This proves the work done by Mezić et al. [300].

The Loop Current and its eddies could intermingle with the oil spill in dual approaches. First, the oil could arrive the Loop Current (at present sensed in minor quantities in May 2010) and move into the Atlantic Ocean drifting toward Europe and Greenland through the Gulf Stream. If the oil were to become trapped in an eddy, this could drive the oil toward Mexico and Texas. Nonetheless, oil trapped within an eddy (Figs. 4.15 and 4.16), circling within the Gulf of Mexico, which caused great damages to the coastline and the marine life.

At the onset of the spill, the LC extended northward to about the latitude of Tampa Bay, and it appeared to be progressing farther north toward the well site. However, the LC began to retreat southward in early May 2010. By mid-May, an anticyclonic LC eddy of 300–400 km diameter developed, accompanied by smaller eddies, and began to separate from the parent LC [298].

In general, the fate of the oil in the ocean depends on many factors, including transport and dispersion by circulation, chemical transformations and biological consumption. Important information, such as the oil flow rate, was unknown throughout the event. Mitigation activities (dispersants, burning and skimming) added further uncertainties [299].

14.12 Why qHop Algorithm can be Used to Simulate and Track the Oil Spill Trajectory Movement?

The trajectory movement of oil spills are produced without noise. In fact, in homogenous oil spill patterns the pixels are connected continuously in which generated neighbor vectors are similar. In this context, the correlation peak corresponds to the oil spill trajectory displacement because the best Grover's search for oil spill individual pixel displacement. It can also be noticed that the quantum Hopfield algorithm delivers current vector variations along the oil spill, which leads to trajectory movements.

Quantum Hopfield is considered a trajectory optimization algorithm. In fact, the main aim of qHop is to acquire a path or trajectory that optimizes a particular amount of interest or any other objective function associated with a certain performance measure, under a set of the given constraints on the dynamics of the system. Therefore, quantum algorithms for the Hopfield network, which can be implemented for automatic pattern recognition of oil spills, reconstruction its shape and optimization of its trajectory movement. In this view, the large data of SAR and surface ocean dynamics can simply encode into the amplitudes of a quantum state. In other words, a quantum implementation (qHop) for the Hopfield network that encodes an exponential number of neurons within the amplitudes of only a polynomially large register of qubits. This complements alternative encodings focusing on a one-to-one correspondence between neurons and qubits.

The main advantage of qHop is involving the threshold of selected large data and also solving incomplete pattern recognition for both oil spill and current pattern. In fact, qHop recognizes the large data as a function of the quantum energy gradient state, which is presented in the eigenbasis of the matrix of multiSAR image energies with high precision. Moreover, the implementation of Grover's search algorithm assists to determine the sharpness pattern of the oil spill trajectory model, which is clear with sharp edges of the mesoscale eddies in Figs. 14.15 to 14.17. In this understanding, Grover's search is considered as the discrete search space, which can be represented by n qubit system. Conversely, Grover operator contains the same amplitude amplification in one iteration, but at the end, the amplitude of the marked item is amplified. In this regard, Grover's iteration should be carried out to acquire the maximum probability of success, which is close to 1. It is worth noting that qHop algorithm is far superior to their classical counterparts, which are of the order of $O(N)$. Quantum Hopfield delivers enhancement pattern of oil spill trajectory movements as compared to previous work carried by Marghany [278] and Cheng et al. [267].

The State-Of-The-Art of oil marine pollution SAR detection and monitoring, as described in this chapter, promises excellent operations of quantum computing algorithms. Yet, there are some concerns that need to be solved. The first involves the revisit time, which is not yet optimized for actual time of operation. In this context, it might be better to reconstruct such constellations with a revisit time of 1 day across the equator, for instance, the fourth Cosmo-SkyMed satellite constellations. Further, the combination of the different SAR sensor observations could be one of the solutions. Indeed, there are plenty of SAR satellite accessibilities, which are increasingly wider with the fourth Cosmo-SkyMed satellite and the two SAOCOM (L-band SAR).

Needless to say, the oil spill thickness cannot be quantified using SAR sensors, which is still accessible matter. Moreover, the chemical characteristics of oil spills and their components also remain a complex task even by using multifrequency and multi polarization SAR data. Indeed, the chemical parameters and thickness are quantified through field measurements, not automatically or estimated in SAR data.

References

[1] Young HD, Freedman RA and Ford L. University Physics Vol 2. Pearson Education; 2007.

[2] Sears FW, Zemansky MW and Young HD. University Physics. Addison-Wesley; 1987.

[3] Young HD, Freedman RA and Ford AL. Sears and Zemansky's University Physics. Pearson Education; 2006 Dec 8.

[4] Linder CJ. University physics students' conceptualizations of factors affecting the speed of sound propagation. International Journal of Science Education. 1993 Nov 1; 15(6): 655–62.

[5] Born M and Wolf E. Principles of Optics: Electromagnetic Theory of Propagation, Interference and Diffraction of Light. Elsevier; 2013 Jun 1.

[6] Born M and Wolf E. Principles of Optics: Electromagnetic Theory of Propagation, Interference and Diffaction of Light. Pergamon Press; 1959.

[7] Ghatak AK. Contemporary Optics. Springer Science & Business Media; 2012 Dec 6.

[8] Ghatak A and Thyagarajan K. An Introduction to Fiber Optics. Cambridge University Press; 1998 Jun 28.

[9] Ghatak A. Optics, McGraw-Hill, New York, 2009.

[10] Griffiths DJ. Resource letter EM-1: Electromagnetic momentum. American Journal of Physics. 2012 Jan; 80(1): 7–18.

[11] Griffiths DJ. Electrodynamics. Introduction to Electrodynamics, 3rd ed., Prentice Hall, Upper Saddle River, New Jersey. 1999: 301–6.

[12] Griffith DJ and Ruppeiner G. Introduction to Electrodynamics. American Journal of Physics. 1981 Dec; 49: 1188–9.

[13] Griffiths DJ. Introduction to Electrodynamics 1981 Prentice Hall. Englewood Cliffs, Nueva Jersey. 1989.

[14] James MB and Griffiths DJ. Why the speed of light is reduced in a transparent medium. American Journal of Physics. 1992 Apr; 60(4): 309–13.

[15] Wheeler JA and Feynman RP. Classical electrodynamics in terms of direct interparticle action. Reviews of Modern Physics. 1949 Jul 1; 21(3): 425.

[16] Hehl FW and Obukhov YN. Foundations of classical electrodynamics: Charge, flux, and metric. Springer Science & Business Media; 2012 Dec 6.

[17] Kumar A and Ghatak AK. Polarization of light with applications in optical fibers. SPIE press; 2011 Jan 25.

[18] Johnson SG and Joannopoulos JD. Block-iterative frequency-domain methods for Maxwell's equations in a planewave basis. Optics express. 2001 Jan 29; 8(3): 173–90.

[19] Zhang Z and Satpathy S. Electromagnetic wave propagation in periodic structures: Bloch wave solution of Maxwell's equations. Physical Review Letters. 1990 Nov 19; 65(21): 2650.

[20] Taflove A and Hagness SC. Computational Electrodynamics: The Finite-Difference Time-Domain Method. Artech House; 2005.

[21] Marghany M. Advanced Remote Sensing Technology for Tsunami Modelling and Forecasting. Boca Raton: CRC Press; 2018.

[22] Dirac PA. The Principles of Quantum Mechanics. Oxford University Press; 1981.

[23] Raymer MG and Smith BJ. The Maxwell wave function of the photon. pp. 293–298. In: The Nature of Light: What Is a Photon? 2005 Aug 4 (Vol. 5866). International Society for Optics and Photonics.

[24] Smith BJ, Killett B, Raymer MG, Walmsley IA and Banaszek K. Measurement of the transverse spatial quantum state of light at the single-photon level. Optics Letters. 2005 Dec 15; 30(24): 3365–7.

[25] Mukamel E, Banaszek K, Walmsley IA and Dorrer C. Direct measurement of the spatial Wigner function with area-integrated detection. Optics Letters. 2003 Aug 1; 28(15): 1317–9.

[26] Pedrosa IA. Quantum description of electromagnetic waves in time-dependent linear media. InJournal of Physics: Conference Series 2011 (Vol. 306, No. 1, p. 012074). IOP Publishing.

[27] de Lima AL, Rosas A and Pedrosa IA. On the quantization of the electromagnetic field in conducting media. Journal of Modern Optics. 2009 Jan 10; 56(1): 41–7.

[28] Pedrosa IA. Exact wave functions of a harmonic oscillator with time-dependent mass and frequency. Physical Review A. 1997 Apr 1; 55(4): 3219.

[29] Alicki R and Kryszewski S. Completely positive Bloch-Boltzmann equations. Physical Review A. 2003 Jul 11; 68(1): 013809.

[30] Ünal N. Quasi-coherent states for a photon in time varying dielectric media. Annals of Physics. 2012 Sep 1; 327(9): 2177–83.

[31] Pedrosa IA, Rosas A and Guedes I. Exact quantum motion of a particle trapped by oscillating fields. Journal of Physics A: Mathematical and General. 2005 Aug 16; 38(35): 7757.

[32] Hasan MZ and Kane CL. Colloquium: topological insulators. Reviews of Modern Physics. 2010 Nov 8; 82(4): 3045.

[33] Shun-Qing. Shen. Topological Insulators: Dirac Equation in Condensed Matter. Springer; 2017.

[34] Fu L, Kane CL and Mele EJ. Topological insulators in three dimensions. Physical Review Letters. 2007 Mar 7; 98(10): 106803.

[35] Devoret MH and Schoelkopf RJ. Superconducting circuits for quantum information: an outlook. Science. 2013 Mar 8; 339(6124): 1169–74.

[36] Newrock R. What are Josephson junctions? How do they work?From https://www.scientificamerican. com/article/what-are-josephson-juncti/.

[37] Tarasov M, Stepantsov E, Lindström T, Kalabukhov A, Ivanov Z and Claeson T. Antenna-coupled planar arrays of Josephson junctions. Physica C: Superconductivity. 2002 Aug 1; 372: 355–9.

[38] Roy A and Devoret M. Introduction to parametric amplification of quantum signals with Josephson circuits. Comptes Rendus Physique. 2016 Aug 1; 17(7): 740–55.

[39] Bengtsson I and Życzkowski K. Geometry of Quantum States: An Introduction to Quantum Entanglement. Cambridge University Press; 2017 Aug 18.

[40] Mallat S. A Wavelet Tour of Signal Processing. Elsevier; 1999 Sep 14.

[41] De Palma G, Trevisan D and Giovannetti V. Gaussian states minimize the output entropy of one-mode quantum Gaussian channels. Physical Review Letters. 2017 Apr 21; 118(16): 160503.

[42] Koch J, Terri MY, Gambetta J, Houck AA, Schuster DI, Majer J, Blais A, Devoret MH, Girvin SM and Schoelkopf RJ. Charge-insensitive qubit design derived from the Cooper pair box. Physical Review A. 2007 Oct 12; 76(4): 042319.

[43] Passerini F and Severini S. Quantifying complexity in networks: the von Neumann entropy. International Journal of Agent Technologies and Systems (IJATS). 2009 Oct 1; 1(4): 58–67.

[44] Rajagopal AK. Von Neumann entropy associated with the haldane exclusion statistics. Physical Review Letters. 1995 Feb 13; 74(7): 1048.

[45] Jozsa R and Schlienz J. Distinguishability of states and von Neumann entropy. Physical Review A. 2000 Jun 5; 62(1): 012301.

[46] Mallat S. A wavelet tour of signal processing. Elsevier; 1999 Sep 14.

[47] Rosso OA, Blanco S, Yordanova J, Kolev V, Figliola A, Schürmann M and Başar E. Wavelet entropy: a new tool for analysis of short duration brain electrical signals. Journal of Neuroscience Methods. 2001 Jan 30; 105(1): 65–75.

[48] Manucharyan VE, Boaknin E, Metcalfe M, Vijay R, Siddiqi I and Devoret M. Microwave bifurcation of a Josephson junction: Embedding-circuit requirements. Physical Review B. 2007 Jul 26; 76(1): 014524.

[49] Nigg SE, Paik H, Vlastakis B, Kirchmair G, Shankar S, Frunzio L, Devoret MH, Schoelkopf RJ and Girvin SM. Black-box superconducting circuit quantization. Physical Review Letters. 2012 Jun 12; 108(24): 240502.

[50] Siddiqi I, Vijay R, Pierre F, Wilson CM, Metcalfe M, Rigetti C, Frunzio L and Devoret MH. RF-driven Josephson bifurcation amplifier for quantum measurement. Physical Review Letters. 2004 Nov 10; 93(20): 207002.

[51] Govenius J, Lake RE, Tan KY and Möttönen M. Detection of zeptojoule microwave pulses using electrothermal feedback in proximity-induced Josephson junctions. Physical Review Letters. 2016 Jul 15; 117(3): 030802.

[52] Hofheinz M, Wang H, Ansmann M, Bialczak RC, Lucero E, Neeley M, O'connell AD, Sank D, Wenner J, Martinis JM and Cleland AN. Synthesizing arbitrary quantum states in a superconducting resonator. Nature. 2009 May; 459(7246): 546.

[53] Bai L, Harder M, Chen YP, Fan X, Xiao JQ and Hu CM. Spin pumping in electrodynamically coupled magnon-photon systems. Physical Review Letters. 2015 Jun 1; 114(22): 227201.

[54] Tserkovnyak Y, Brataas A and Bauer GE. Enhanced Gilbert damping in thin ferromagnetic films. Physical Review Letters. 2002 Feb 28; 88(11): 117601.

[55] Mosendz O, Pearson JE, Fradin FY, Bauer GE, Bader SD and Hoffmann A. Quantifying spin Hall angles from spin pumping: Experiments and theory. Physical Review Letters. 2010 Jan 28; 104(4): 046601.

[56] Huebl H and Goennenwein ST. Electrical signal picks up a magnet's heartbeat. Physics. 2015 Jun 1; 8: 51.

[57] Matthew JA. Spin dependence of the electron inelastic mean free path and the elastic scattering cross section—a high-energy atomic approximation. Physical Review B. 1982 Mar 1; 25(5): 3326.

[58] Gilmore R. Elementary quantum mechanics in one dimension. JHU Press; 2004 Sep 17.

[59] Ter Haar D, editor. Problems in quantum mechanics. Courier Corporation; 2014 Jun 10.

[60] Valavanis A, Ikonić Z and Kelsall RW. Intersubband carrier scattering in n-and p− Si∕Si Ge quantum wells with diffuse interfaces. Physical Review B. 2008 Feb 11; 77(7): 075312.

[61] Dirac PA. The Lagrangian in quantum mechanics. pp. 111–119 *In*: Feynman's Thesis—A New Approach To Quantum Theory 2005.

[62] Chandler D and Wolynes PG. Exploiting the isomorphism between quantum theory and classical statistical mechanics of polyatomic fluids. The Journal of Chemical Physics. 1981 Apr 1; 74(7): 4078–95.

[63] Schwinger J. Quantum Kinematics and Dynamic. CRC Press; 2018 Mar 5.

[64] Baym G. Lectures on Quantum Mechanics. CRC Press; 2018 Mar 5.

[65] Shortell MP, Hewins RA, Fernando JF, Walden SL, Waclawik ER and Jaatinen EA. Multi-angle fluorometer technique for the determination of absorption and scattering coefficients of subwavelength nanoparticles. Optics Express. 2016 Jul 25; 24(15):17090–102.

[66] Karam MA, Amar F, Fung AK, Mougin E, Lopes A, Le Vine DM and Beaudoin A. A microwave polarimetric scattering model for forest canopies based on vector radiative transfer theory. Remote Sensing of Environment. 1995 Jul 1; 53(1): 16–30.

[67] Ulaby FT., Moore RK and Fung AK. Microwave Remote Sensing, Active and Passive, Artech House, Inc, 1986.

[68] Cohen-Tannoudji C, Dupont-Roc J and Grynberg G. Atom-photon interactions: basic processes and applications. Atom-Photon Interactions: Basic Processes and Applications, by Claude Cohen-Tannoudji, Jacques Dupont-Roc, Gilbert Grynberg, pp. 678. ISBN 0-471-29336-9. Wiley-VCH, March 1998.1998 Mar: 678.

[69] Ryder LH. Quantum field theory. Cambridge University Press; 1996 Jun 6.

[70] Weinberg S. The Quantum Theory of Fields I. Cambridge University Press, 1995.

[71] Sullivan R. Radar foundations for imaging and advanced concepts. The Institution of Engineering and Technology; 2004.

[72] Ahern FJ. Fundamental concepts of imaging radar: basic level; unpublished manual, Canada Centre for Remote Sensing, Ottawa, Ontario, 87 p, 1995.

[73] Bamler R. Principles of synthetic aperture radar. Surveys in Geophysics. 2000 Mar 1; 21(2-3): 147–57.

[74] Moreira A. Real-time synthetic aperture radar (SAR) processing with a new subaperture approach. IEEE Transactions on Geoscience and Remote Sensing. 1992 Jul; 30(4): 714–22.

[75] Lloyd S. Enhanced sensitivity of photodetection via quantum illumination. Science. 2008 Sep 12; 321(5895): 1463–5.

[76] Lacomme P. Air and spaceborne radar systems. William Andrew; 2001.

[77] Merrill IS. Introduction to radar systems. Mc Grow-Hill. 2001: 607–9.

[78] Skolnik ML. *Radar Handbook*, Third Edition, McGraw Hill, 2008.

[79] McAulay AD. Military laser technology for defense: Technology for revolutionizing 21st century warfare. John Wiley & Sons; 2011 Apr 20.

[80] Lanzagorta M. Quantum radar. Synthesis Lectures on Quantum Computing. 2011 Oct 31; 3(1): 1–39.

[81] Smith JF. Quantum entangled radar theory and a correction method for the effects of the atmosphere on entanglement. In Quantum Information and Computation VII 2009 Apr 28 (Vol. 7342, p. 73420A). International Society for Optics and Photonics.

[82] Yao XW, Wang H, Liao Z, Chen MC, Pan J, Li J, Zhang K, Lin X, Wang Z, Luo Z and Zheng W. Quantum Image Processing and Its Application to Edge Detection: Theory and Experiment. Physical Review X. 2017 Sep 11; 7(3): 031041.

[83] Johnson RC, Jasik H. Antenna engineering handbook. New York, McGraw-Hill Book Company, 1984, 1356 p.

[84] Balanis CA. Antenna Theory: Analysis and Design. John Wiley & Sons; 2016.

[85] Volakis JL. Antenna engineering handbook. Johnson RC and Jasik H (editors). New York: McGraw-Hill; 2007 Jun 7.

[86] Widrow B, Mantey PE, Griffiths LJ and Goode BB. Adaptive antenna systems. Proceedings of the IEEE. 1967 Dec; 55(12): 2143–59.

[87] Soumekh M. Synthetic aperture radar signal processing. New York: Wiley; 1999.

[89] Scheer J, Holm WA. Principles of modern radar. Richards MA and Melvin WL (editors). SciTech Pub.; 2010 May.

[90] Shamsoddini A and Trinder JC. Edge-detection-based filter for SAR speckle noise reduction. International Journal of Remote Sensing. 2012 Apr 10; 33(7): 2296–320.

[91] Bhattarai R. Land Subsidence Mapping and Risk Assessment in Kathmandu Valley, Nepal, using D-InSAR and GIS Techniques (Doctoral dissertation, 千 大学= Chiba University).

[92] Moreira A. Real-time synthetic aperture radar (SAR) processing with a new subaperture approach. IEEE Transactions on Geoscience and Remote sensing. 1992 Jul; 30(4): 714–22.

[93] Currie A and Brown MA. Wide-swath SAR. In IEE Proceedings F-Radar and Signal Processing 1992 Apr (Vol. 139, No. 2, pp. 122–135). IET.

[94] Bamler R and Eineder M. Accuracy of differential shift estimation by correlation and split-bandwidth interferometry for wideband and delta-k SAR systems. IEEE Geoscience and Remote Sensing Letters. 2005 Apr; 2(2): 151–5.

[95] Currie A and Brown MA. Wide-swath SAR. In IEE Proceedings F-Radar and Signal Processing 1992 Apr (Vol. 139, No. 2, pp. 122–135). IET.

[96] Zhang S, Long T, Zeng T and Ding Z. Space-borne synthetic aperture radar received data simulation based on airborne SAR image data. Advances in Space Research. 2008 Jan 1; 41(11): 1818–21.

[97] Ramirez AB, Rivera IJ and Rodriguez D. Sar image processing algorithms based on the ambiguity function. InCircuits and Systems, 2005. 48th Midwest Symposium on 2005 Aug 7 (pp. 1430–1433). IEEE.

[98] Laur H, Bally P, Meadows P, Sanchez J, Schaettler B, Lopinto E and Esteban D. Derivation of the backscattering coefficient σ0 in ESA ERS SAR PRI products. ESA, Noordjiwk, The Netherlands, ESA Document ES-TN-RE-PM-HL09. 2002 Sep(2).

[99] Lu Z and Meyer DJ. Study of high SAR backscattering caused by an increase of soil moisture over a sparsely vegetated area: implications for characteristics of backscattering. International Journal of Remote Sensing. 2002 Jan 1; 23(6): 1063–74.

[100] Budge MC and Burt MP. Range correlation effects in radars. In The Record of the 1993 IEEE National Radar Conference 1993 (pp. 212–216). IEEE.

[101] Budge MC and Gilbert SM. Timing jitter spectrum in pulsed and pulsed Doppler radars. In Proceedings of Southeastc on'93 1993 Apr 4 (pp. 4-p). IEEE.

[102] Budge MC and German SR. Basic RADAR analysis. Artech House; 2015 Oct 1.

[103] Fortuny J and Sieber AJ. Fast algorithm for a near-field synthetic aperture radar processor. IEEE Transactions on Antennas and Propagation. 1994 Oct; 42(10): 1458–60.

[104] Werness SA, Carrara WG, Joyce LS and Franczak DB. Moving target imaging algorithm for SAR data. IEEE Transactions on Aerospace and Electronic Systems. 1990 Jan; 26(1): 57–67.

[105] Bamler R. A comparison of range-Doppler and wavenumber domain SAR focusing algorithms. IEEE Transactions on Geoscience and Remote Sensing. 1992 Jul; 30(4): 706–13.

[106] Yigit E, Demirci S, Ozdemir C and Kavak A. A synthetic aperture radar-based focusing algorithm for B-scan ground penetrating radar imagery. Microwave and Optical Technology Letters. 2007 Oct; 49(10): 2534–40.

[107] Zhu D, Li Y and Zhu Z. A keystone transform without interpolation for SAR ground moving-target imaging. IEEE Geoscience and Remote Sensing Letters. 2007 Jan; 4(1): 18–22.

[108] Stepinski T. An implementation of synthetic aperture focusing technique in the frequency domain. IEEE transactions on ultrasonics, ferroelectrics, and frequency control. 2007 Jul; 54(7): 1399–408.

[109] Kingsley S and Quegan S. Understanding radar systems. SciTech Publishing; 1999.

[110] Lillesand T, Kiefer RW and Chipman J. Remote sensing and image interpretation. John Wiley & Sons; 2014 Dec 31.

[111] Ahern FJ. Fundamental Concepts of Imaging Radar: Basic Level. Unpublished manual, Canada Centre for Remote Sensing, Ottawa. 1995.

[112] Thomas ML, Ralph WK and Jonathan WC. Remote sensing and image interpretation. John Willey & Sons, New York. 2000.

[113] Shi J and Dozier J. Measurements of snow-and glacier-covered areas with single-polarization SAR. Annals of Glaciology. 1993 Jan; 17: 72–6.

[114] Zhang WG, Zhang Q and Yang CS. Improved bilateral filtering for SAR image despeckling. Electronics Letters. 2011 Feb 17; 47(4): 286–8.

[115] UNAVACO. Synthetic Aperture Radar (SAR) Satellites. https://www.unavco.org/instrumentation/geophysical/imaging/sar-satellites/sar-satellites.html. Access on 19 January 2019.

[116] Brisco B. Mapping and monitoring surface water and wetlands with synthetic aperture radar. Remote Sensing of Wetlands: Applications and Advances. 2015 Mar 23: 119–36.

[117] Headrick JM, Anderson SJ and Skolnik M. HF over-the-horizon radar. Radar Handbook. 2008 Feb; 3.

[118] Skolnik M, Linde G and Meads K. Senrad: An advanced wideband air-surveillance radar. IEEE Transactions on aerospace and electronic systems. 2001 Oct; 37(4): 1163–75.

[119] Gardner JP, Sharpies RM, Carrasco BE and Frenk CS. A wide-field K-band survey—I. Galaxy counts in B, V, I and K. Monthly Notices of the Royal Astronomical Society. 1996 Sep 1; 282(1): L1-6.

[120] Carman E, Case M, Kamegawa M, Yu R, Giboney K and Rodwell M. V-band and W-band broadband, monolithic distributed frequency multipliers. In 1992 IEEE MTT-S Microwave Symposium Digest 1992 Jun 1 (pp. 819–822). IEEE.

[121] Kunz WH and Hildebrandt PR. Inventors; US Air Force, assignee. V-band coupling. United States patent US 3,861,723. 1975 Jan 21.

[122] Gjurchinovski A. Reflection from a moving mirror—a simple derivation using the photon model of light. European Journal of Physics. 2012 Nov 28; 34(1): L1.

[123] Anglada-Escudé G, Klioner SA, Soffel M, Torra J. Relativistic effects on imaging by a rotating optical system. Astronomy & Astrophysics. 2007 Jan 1; 462(1): 371–7.

[124] Gjurchinovski A. Reflection of light from a uniformly moving mirror. American journal of physics. 2004 Oct; 72(10): 1316–24.

[125] Gjurchinovski A. Einstein's mirror and Fermat's principle of least time. American Journal of Physics. 2004 Oct; 72(10): 1325–7.

[126] Gjurchinovski A. The Doppler effect from a uniformly moving mirror. European Journal of Physics. 2005 May 20; 26(4): 643.

[127] Ashby N. Relativity in the global positioning system. Living Reviews in Relativity. 2003 Dec 1; 6(1): 1.

[128] Espedal H. Detection of oil spill and natural film in the marine environment by spaceborne SAR. In IEEE 1999 International Geoscience and Remote Sensing Symposium. IGARSS'99 (Cat. No. 99CH36293) 1999 (Vol. 3, pp. 1478–1480). IEEE.

[129] Hovland HA, Johannessen JA, Digranes G. Slick detection in SAR images. In Proceedings of IGARSS'94-1994 IEEE International Geoscience and Remote Sensing Symposium 1994 Aug 8 (Vol. 4, pp. 2038–2040). IEEE.

[130] Fingas MF and Brown CE. Review of oil spill remote sensing. Spill Science & Technology Bulletin. 1997 Jan 1; 4(4): 199–208.

[131] MARTiNEZ AN and Moreno V. An oil spill monitoring system based on SAR images. Spill science & technology bulletin. 1996 Jan 1; 3(1-2): 65–71.

[132] Vespe M and Greidanus H. SAR image quality assessment and indicators for vessel and oil spill detection. IEEE Transactions on Geoscience and Remote Sensing. 2012 Nov; 50(11): 4726–34.

[133] Bondi H. Assumption and myth in physical theory: the tarner lectures delivered at Cambridge in November 1965. CUP Archive; 1967.

[134] Dolby CE and Gull SF. On radar time and the twin "paradox". American Journal of Physics. 2001 Dec; 69(12): 1257–61.

[135] Shapiro II, Ash ME, Ingalls RP, Smith WB, Campbell DB, Dyce RB, Jurgens RF and Pettengill GH. Fourth test of general relativity: new radar result. Physical Review Letters. 1971 May 3; 26(18): 1132.

[136] Bunkin BV, Gaponov-Grekhov AV, El'Chaninov AS, Zagulov FI, Korovin SD, Mesiats GA, Osipov ML, Otlivanchik EA, Petelin MI and Prokhorov AM. Radar-based on a microwave oscillator with a relativistic electron beam. Pisma v Zhurnal Tekhnischeskoi Fiziki. 1992 May; 18: 61–5.

[137] Krieger G and De Zan F. Relativistic effects in bistatic SAR processing and system synchronization. In EUSAR 2012; 9th European Conference on Synthetic Aperture Radar 2012 Apr 23 (pp. 231–234). VDE.

[138] Williams MO. Quantum mechanics of hydrocarbon chains,from users.physics.harvard. edu/~mwilliams/.../Quantum-Mechanics-of-Hydrocarbons.pdf;2011 Apr. 3.

[139] Peres A. Quantum theory: concepts and methods. Springer Science & Business Media; 2006 Jun 1.

[140] Namias V. The fractional order Fourier transform and its application to quantum mechanics. IMA Journal of Applied Mathematics. 1980 Mar 1; 25(3): 241–65.

[141] Romero FD, Pitcher MJ, Hiley CI, Whitehead GF, Kar S, Ganin AY, Antypov D, Collins C, Dyer MS, Klupp G and Colman RH. Redox-controlled potassium intercalation into two polyaromatic hydrocarbon solids. Nature Chemistry. 2017 Jul; 9(7): 644.

[141] Basdevant JL and Dalibard J. The quantum mechanics solver: how to apply quantum theory to modern physics. (Springer-Verlag, Berlin Heidelberg 2006) Phys. Rev. pp. 231–35.

[142] Lowe JP and Peterson K. Quantum Chemistry. Elsevier; 2011 Aug 30.

[143] Downare TD and Mullins OC. Visible and near-infrared fluorescence of crude oils. Applied Spectroscopy. 1995 Jun; 49(6): 754–64.

[144] Junli Han, The research for infrared image target recognition technology Master Thesis, Nanjing University of Science and Technology, 2004.

[145] Verboven E. On the quantum theory of electrical conductivity: The conductivity tensor to zeroth order. Physica. 1960 Dec 1; 26(12): 1091–116.

[146] Nakajima S. On quantum theory of transport phenomena: Steady diffusion. Progress of Theoretical Physics. 1958 Dec 1; 20(6): 948–59.

[147] Mott NF. The electrical conductivity of transition metals. Proc. R. Soc. Lond. A. 1936 Feb 1; 153(880): 699–717.

[148] Kohn W and Luttinger JM. Quantum theory of electrical transport phenomena. Physical Review. 1957 Nov 1; 108(3): 590.

[149] Adams EN and Holstein TD. Quantum theory of transverse galvano-magnetic phenomena. Journal of Physics and Chemistry of Solids. 1959 Aug 1; 10(4): 254–76.

[150] Plant WJ and Keller WC. Evidence of Bragg scattering in microwave Doppler spectra of sea return. Journal of Geophysical Research: Oceans. 1990 Sep 15; 95(C9): 16299–310.

[151] Plant WJ. Bragg scattering of electromagnetic waves from the air/sea interface. In Surface Waves and Fluxes 1990 (pp. 41–108). Springer, Dordrecht.

[152] Wright JW. A new model for sea clutter. IEEE Transactions on Antennas and Propagation. 1968 Mar; 16(2): 217–23.

[153] Gade M, Alpers W, Hühnerfuss H, Masuko H and Kobayashi T. Imaging of biogenic and anthropogenic ocean surface films by the multifrequency/multipolarization SIR-C/X-SAR. Journal of Geophysical Research: Oceans. 1998 Aug 15; 103(C9): 18851–66.

[154] Liu K, Xiao H, Fan H and Fu Q. Analysis of quantum radar cross section and its influence on target detection performance. IEEE Photonics Technology Letters. 2014 Jun 1; 26(11): 1146–9.

[155] Fang C, Chen Y, Xu Y and Hua L. The Analysis of Change Factor of the Simulation of the Bistatic Quantum Radar Cross Section for the Typical Ship Structure. In 2018 IEEE Asia-Pacific Conference on Antennas and Propagation (APCAP) 2018 Aug 5 (pp. 190–193). IEEE.

[156] Tegmark M. Apparent wave function collapse caused by scattering. Foundations of Physics Letters. 1993 Dec 1; 6(6): 571–90.

[157] Migliaccio M, Nunziata F and Gambardella A. On the co-polarized phase difference for oil spill observation. International Journal of Remote Sensing. 2009 Mar 1; 30(6): 1587–602.

[158] Ulaby FT, Kouyate F, Brisco B and Williams TL. Textural information in SAR images. IEEE Transactions on Geoscience and Remote Sensing. 1986 Mar (2): 235–45.

[159] Wei L, Hu Z, Guo M, Jiang M and Zhang S. Texture feature analysis in oil spill monitoring by SAR image. In 2012 20th International Conference on Geoinformatics 2012 Jun 15 (pp. 1–6). IEEE.

[160] Haralick RM and Shanmugam K. Textural features for image classification. IEEE Transactions on systems, man, and cybernetics. 1973 Nov(6): 610–21.

[161] Zhang F, Shao Y, Tian W and Wang S. Oil spill identification based on textural information of SAR image. In IGARSS 2008-2008 IEEE International Geoscience and Remote Sensing Symposium 2008 Jul 7 (Vol. 4, pp. IV-1308). IEEE.

[162] Hall-Beyer M, GLCM texture: A tutorial Version 2.3.; Department of Geography, University of Calgary: Calgary, Alberta, Canada, 2000. http://www.fp.ucalgary.ca/mhallbey/, 2019.Maged

[163] Marghany M. RADARSAT automatic algorithms for detecting coastal oil spill pollution. International Journal of Applied Earth Observation and Geoinformation. 2001 Jan 1; 3(2): 191–6.

[164] Peckinpaugh SH. An improved method for computing gray-level cooccurrence matrix based texture measures. CVGIP: Graphical Models and Image Processing. 1991 Nov 1; 53(6): 574–80.

[165] Robert MK. Textural features for image classification. IEEE Transactions on Sytems, Man, and Cybernetics. 1973; 3(6): 610–21.

[166] Weszka JS, Dyer CR and Rosenfeld A. A comparative study of texture measures for terrain classification. IEEE transactions on Systems, Man, and Cybernetics. 1976 Apr(4): 269–85.

[167] Wu CM and Chen YC. Statistical feature matrix for texture analysis. CVGIP: Graphical Models and Image Processing. 1992 Sep 1; 54(5): 407–19.

[168] Conners RW and Harlow CA. A theoretical comparison of texture algorithms. IEEE transactions on pattern analysis and machine intelligence. 1980 May (3): 204–22.

[169] Le Hur K. Entanglement entropy, decoherence, and quantum phase transitions of a dissipative two-level system. Annals of Physics. 2008 Sep 1; 323(9): 2208–40.

[170] Topouzelis K. Oil spill detection by SAR images: dark formation detection, feature extraction and classification algorithms. Sensors. 2008 Oct; 8(10): 6642–59.

[171] Del Frate F, Petrocchi A, Lichtenegger J and Calabresi G. Neural networks for oil spill detection using ERS-SAR data. IEEE Transactions on geoscience and remote sensing. 2000 Sep; 38(5): 2282–7.

[172] Topouzelis K, Karathanassi V, Pavlakis P and Rokos D. Detection and discrimination between oil spills and look-alike phenomena through neural networks. ISPRS Journal of Photogrammetry and Remote Sensing. 2007 Sep 1; 62(4): 264–70.

[173] Topouzelis K, Karathanassi V, Pavlakis P and Rokos D. Potentiality of feed-forward neural networks for classifying dark formations to oil spills and look-alikes. Geocarto International. 2009 Jun 1; 24(3): 179–91.

[174] Garcia-Pineda O, Macdonald I, Hu C, Svejkovsky J, Hess M, Dukhovskoy D and Morey SL. Detection of floating oil anomalies from the Deepwater Horizon oil spill with synthetic aperture radar. Oceanography. 2013 Jun 1; 26(2): 124–37.

[175] Li Y and Zhang Y. Synthetic aperture radar oil spills detection based on morphological characteristics. Geo-spatial Information Science. 2014 Jan 2; 17(1): 8–16.

[176] Solberg AS, Storvik G, Solberg R and Volden E. Automatic detection of oil spills in ERS SAR images. IEEE Transactions on Geoscience and remote sensing. 1999 Jul; 37(4): 1916–24.

[177] Fiscella B, Giancaspro A, Nirchio F, Pavese P and Trivero P. Oil spill detection using marine SAR images. International Journal of Remote Sensing. 2000 Jan 1; 21(18): 3561–6.

[178] Gil P and Alacid B. Oil Spill Detection in Terma-Side-Looking Airborne Radar Images Using Image Features and Region Segmentation. Sensors. 2018 Jan; 18(1): 151.

[179] Mohamed IS, Salleh AM and Tze LC. Detection of oil spills in Malaysian waters from RADARSAT Synthetic Aperture Radar data and prediction of oil spill movement. In Proc. of 19th Asi. Conf. on Rem. Sen., Hong Kong, China, November 1999 Nov (pp. 23–27).

[180] Samad R and Mansor SB. Detection of oil spill pollution using RADARSAT SAR imagery. InCD Proc. of 23rd Asi. Conf. on Rem. Sens. Birendra International Convention Centre in Kathmandu, Nepal, November 2002 Nov 25 (pp. 25–29).

[181] De Maesschalck R, Jouan-Rimbaud D and Massart DL. The mahalanobis distance. Chemometrics and Intelligent Laboratory Systems. 2000 Jan 4; 50(1): 1–8.

[182] Negnevitsky M and Intelligence A. A guide to intelligent systems. Artificial Intelligence, 2nd edition, pearson Education. 2005.

[183] Haykin S. Neural Networks, A comprehensive Foundation Second Edition by Prentice-Hall.1999.

[184] Marghany M, Cracknell AP and Hashim M. Modification of fractal algorithm for oil spill detection from RADARSAT-1 SAR data. International Journal of Applied Earth Observation and Geoinformation. 2009 Apr 1; 11(2): 96–102.

[185] Marghany M, Hashim M and Cracknell AP. Fractal dimension algorithm for detecting oil spills using RADARSAT-1 SAR. InInternational Conference on Computational Science and Its Applications 2007 Aug 26 (pp. 1054–1062). Springer, Berlin, Heidelberg.

[186] Bishop CM. Neural networks for pattern recognition: Oxford University Press. New York. 1996.

[187] Mandelbrot BB. The fractal geometry of nature. New York: WH freeman; 1983 Mar 1.

[188] Falconer K. Fractal geometry: mathematical foundations and applications. John Wiley & Sons; 2004 Jan 9.

[189] Briggs J. Fractals: The patterns of chaos: A new aesthetic of art, science, and nature. Simon and Schuster; 1992.

[190] Nittmann J, Daccord G and Stanley HE. Fractal growth viscous fingers: quantitative characterization of a fluid instability phenomenon. Nature. 1985 Mar; 314(6007): 141.

[191] Pietronero L and Tosatti E, editors. Fractals in physics. Elsevier; 2012 Dec 2.

[192] Edgar G. Measure, topology, and fractal geometry. Springer Science & Business Media; 2007 Oct 23.

[193] Balay-Karperien A. Defining Microglial Morphology: Form, Function and Fractal Dimension (Doctoral dissertation, Charles Sturt University); 2004.

[194] Redondo JM. Fractal Description of density interfaces. Journal of Mathematics and its Applications. 1996 May 5: 210–218.

[195] Pentland AP. Fractal-based description of natural scenes. IEEE Transactions on Pattern Analysis & Machine Intelligence. 1984 Jun 1(6): 661–74.

[196] Falconer K. Fractal Geometry John Wiley & Sons. Inc., Chichester. 1990.

[197] Benelli G, Garzelli A. Oil-spills detection in SAR images by fractal dimension estimation. InIEEE 1999 International Geoscience and Remote Sensing Symposium. IGARSS'99 (Cat. No. 99CH36293) 1999 (Vol. 1, pp. 218–220). IEEE.

[198] Marghany M, Cracknell AP and Hashim M. Comparison between radarsat-1 SAR different data modes for oil spill detection by a fractal box counting algorithm. International Journal of Digital Earth. 2009 Sep 1; 2(3): 237–56.

[199] Milan S, Vachav H and Roger B. Image Processing Analysis and Machine Vision. Chapman and Hall Computing, New York.1993.

[200] Sarkar N and Chaudhuri BB. An efficient differential box-counting approach to compute fractal dimension of image. IEEE Transactions on systems, man, and cybernetics. 1994 Jan; 24(1): 115–20.

[201] Vala HJ and Baxi A. A review on Otsu image segmentation algorithm. International Journal of Advanced Research in Computer Engineering & Technology (IJARCET). 2013 Feb 28; 2(2): 387–9.

[202] Huang M, Yu W and Zhu D. An improved image segmentation algorithm based on the Otsu method. In 2012 13th ACIS International Conference on Software Engineering, Artificial Intelligence, Networking and Parallel/Distributed Computing 2012 Aug 8 (pp. 135–139). IEEE.

[203] Annis AA and Lloyd EH. The expected value of the adjusted rescaled Hurst range of independent normal summands. Biometrika. 1976 Jan 1; 63(1): 111–6.

[204] Kotova L, Espedal HA and Johannessen OM. Oil spill detection using spaceborne SAR: a brief review. Proceedings of 27th International Symposium on Remote Sensing Environmental, 8–12 June 1998, Norwegian Defence Research Establishment. Tromsø, Norway, 1998 pp. 791–794.

[205] Ivanov A, He M and Fang MQ. Oil spill detection with the RADARSAT SAR in the waters of the Yellow and East Sea: A case study CD of 23rd Asian Conference on Remote Sensing, 13–17 November 2002, Nepal, Asian Remote Sensing Society, Japan. 2002 Vol 1, pp. 1–8.

[206] Marghany M, Cracknell AP and Hashim M. Modification of fractal algorithm for oil spill detection from RADARSAT-1 SAR data. International Journal of Applied Earth Observation and Geoinformation. 2009 Apr 1; 11(2): 96–102.

[207] Bertacca M, Berizzi F and Mese ED. A FARIMA-based technique for oil slick and low-wind areas discrimination in sea SAR imagery. IEEE transactions on geoscience and remote sensing. 2005 Nov; 43(11): 2484–93.

[208] Gade M and Redondo JM. Marine pollution in european coastal waters monitored by the ERS-2 SAR: a comprehensive statistical analysis. In Proceedings of Geoscience and Remote Sensing Symposium, 1999, IGARSS'99, Hamburg, Germany, 28 June–2 July 1999, IEEE Geoscience and Remote Sensing Society, USA. 1999 Vol 2, pp. 1375–1377.

[209] Azghadi MR, Kavehie O and Navi K. A novel design for quantum-dot cellular automata cells and full adders. arXiv preprint arXiv:1204.2048. 2012 Apr 10.

[210] Amlani I, Orlov AO, Kummamuru RK, Bernstein GH, Lent CS and Snider GL. Experimental demonstration of a leadless quantum-dot cellular automata cell. Applied Physics Letters. 2000 Jul 31; 77(5): 738–40.

[211] Orlov AO, Amlani I, Kummamuru RK, Ramasubramaniam R, Toth G, Lent CS, Bernstein GH, Snider GL. Experimental demonstration of clocked single-electron switching in quantum-dot cellular automata. Applied Physics Letters. 2000 Jul 10; 77(2): 295–7.

[212] Bernstein GH, Amlani I, Orlov AO, Lent CS and Snider GL. Observation of switching in a quantum-dot cellular automata cell. Nanotechnology. 1999 Jun; 10(2): 166.

[213] Perez-Martinez F, Farrer I, Anderson D, Jones GA, Ritchie DA, Chorley SJ and Smith CG. Demonstration of a quantum cellular automata cell in a Ga As⁄Al Ga As heterostructure. Applied physics letters. 2007 Jul 16; 91(3): 032102.

[214] Hashemi S, Tehrani M and Navi K. An efficient quantum-dot cellular automata full-adder. Scientific Research and Essays. 2012 Jan 16; 7(2): 177–89.

[215] Bonyadi MR, Azghadi SM, Rad NM, Navi K and Afjei E. Logic optimization for majority gate-based nanoelectronic circuits based on genetic algorithm. In 2007 International Conference on Electrical Engineering 2007 Apr 11 (pp. 1–5). IEEE.

[216] Lent CS, Tougaw PD, Porod W and Bernstein GH. Quantum cellular automata. Nanotechnology. 1993 Jan; 4(1): 49.

[217] Lent CS, Tougaw PD and Porod W. Bistable saturation in coupled quantum dots for quantum cellular automata. Applied Physics Letters. 1993 Feb 15; 62(7): 714–6.

[218] Zhang R, Gupta P and Jha NK. Synthesis of majority and minority networks and its applications to QCA, TPL and SET based nanotechnologies. In 18th International Conference on VLSI Design held jointly with 4th International Conference on Embedded Systems Design 2005 Jan 3 (pp. 229–234). IEEE.

[219] Wiesner K. Quantum cellular automata. Cellular Automata: A Volume in the Encyclopedia of Complexity and Systems Science, Second Edition. 2018: 93–104.

[220] Brennen GK and Williams JE. Entanglement dynamics in one-dimensional quantum cellular automata. Physical Review A. 2003 Oct 13; 68(4): 042311.

[221] NOAA OR&R, 2013. Deepwater Horizon trajectory map archive. Web Document <http://archive.orr.noaa.gov>, [Accessed on April 8 2019].

[222] Lynn KS, Benjamin J, Jodi KB, Patrick M, McCaskill EC, Eric U, Frank M, George RH Jr, Ole MS and Patrick H 2011). Airborne Ocean Surveys of the Loop Current Complex From NOAA WP-3D in Support of the Deepwater Horizon Oil Spill, in Monitoring and Modeling the Deepwater Horizon Oil Spill: A Record-Breaking Enterprise. Liu Y, Macfadyen A, Ji Z-G and Weisberg RH (eds.). American Geophysical Union, Washington, D. C., pp. 131–151.

[223] Nan DW, Chet TP, Vandana VR, Eurico JD'Sa, Robert RL, Nicholas GH, Peter JB, Patrice DC, Neha S, Hans CG and Raymond E. 2011. TurnerImpacts of loop current Frontal cyclonic eddies and wind forcing on the 2010 Gulf of Mexico oil spill, in Monitoring and Modeling the Deepwater Horizon Oil Spill: A Record-Breaking Enterprise (eds Y. Liu, A. Macfadyen, Z.-G. Ji and R. H. Weisberg), American Geophysical Union, Washington, D. C., pp. 103–116.

[224] NOAA/NESDIS. 2013. National environmental satellite information service, experimental marine pollution surveillance daily composite product. Digital Archive. <http://satepsanone.nesdis.noaa.gov/OMS/disasters/DeepwaterHorizon/composites/2010/>, [Access on April 8 2019].

[225] Zangari G. Risk of global climate change by BP oil spill. National Inst. of Nuclear Ph., Italy (2010) (unpublished). [www.associazionegeofisica.it/OilSpill.pdf] [Accessed on February 7 2019].

[226] Cheng A, Arkett M, Zagon T, De Abreu R, Mueller D, Vachon P and Wolfe J. Oil detection in RADARSAT-2 quad-polarization imagery: Implications for ScanSAR performance. InSAR Image Analysis, Modeling, and Techniques XI 2011 Oct 27 (Vol. 8179, p. 81790G). International Society for Optics and Photonics.

[227] Caruso MJ, Migliaccio M, Hargrove JT, Garcia-Pineda O and Graber HC. Oil spills and slicks imaged by synthetic aperture radar. Oceanography. 2013 Jun 1; 26(2): 112–23.

[228] Garcia-Pineda O, MacDonald IR, Li X, Jackson CR and Pichel WG. Oil spill mapping and measurement in the Gulf of Mexico with textural classifier neural network algorithm (TCNNA). IEEE Journal of Selected Topics in Applied Earth Observations and Remote Sensing. 2013 Dec; 6(6): 2517–25.

[229] Marghany M. Automatic Mexico Gulf Oil Spill Detection from Radarsat-2 SAR Satellite Data Using Genetic Algorithm. Acta Geophysica. 2016 Oct 1; 64(5): 1916–41.

[230] Lounis B and Belhadj-Aissa A. Sea SAR images analysis to detect oil slicks in Algerian coasts. Journal of Mathematical Modelling and Algorithms in Operations Research. 2014 Dec 1; 13(4): 371–86.

[231] Han KH and Kim JH. Quantum-inspired evolutionary algorithm for a class of combinatorial optimization. IEEE transactions on evolutionary computation. 2002 Dec; 6(6): 580–93.

[232] Jiao L, Li Y, Gong M and Zhang X. Quantum-inspired immune clonal algorithm for global optimization. IEEE Transactions on Systems, Man, and Cybernetics, Part B (Cybernetics). 2008 Oct; 38(5): 1234–53.

[233] Williams CP. Explorations in quantum computing. Springer Science & Business Media; 2010 Dec 7.

[234] Kaye P, Laflamme R and Mosca M. An introduction to quantum computing. Oxford University Press; 2007.

[235] Hey T. Quantum computing: an introduction. Computing & Control Engineering Journal. 1999 Jun 1; 10(3): 105–12.

[236] Warren WS. The usefulness of NMR quantum computing. Science. 1997 Sep 12; 277(5332): 1688–90.

[237] Braunstein SL, Caves CM, Jozsa R, Linden N, Popescu S and Schack R. Separability of very noisy mixed states and implications for NMR quantum computing. Physical Review Letters. 1999 Aug 2; 83(5): 1054.

[238] Platzman PM and Dykman MI. Quantum computing with electrons floating on liquid helium. Science. 1999 Jun 18; 284(5422): 1967–9.

[239] Beige A, Braun D, Tregenna B and Knight PL. Quantum computing using dissipation to remain in a decoherence-free subspace. Physical review letters. 2000 Aug 21; 85(8): 1762.

[240] Rieffel E, Polak W. An introduction to quantum computing for non-physicists. ACM Computing Surveys (CSUR). 2000 Sep 1; 32(3): 300–35.

[241] Wang L, Tang F and Wu H. Hybrid genetic algorithm based on quantum computing for numerical optimization and parameter estimation. Applied Mathematics and Computation. 2005 Dec 15; 171(2): 1141–56.

[242] Zhang G, Jin W and Hu L. Quantum evolutionary algorithm for multi-objective optimization problems. InProceedings of the 2003 IEEE International Symposium on Intelligent Control 2003 Oct 8 (pp. 703–708). IEEE.

[243] Deb K. Nonlinear goal programming using multi-objective genetic algorithms. Journal of the Operational Research Society. 2001 Mar 1; 52(3): 291–302.

[244] Deb K, Agrawal S, Pratap A and Meyarivan T. A Fast Elitist Non-Dominated Sorting Genetic Algorithm for Multi-Objective Optimization: NSGA-II; In Parallel Problem Solving from Nature-PPSN VI, Proceeding of 6th International Conference, Paris, France, 18–20 September 2000; Schoenauer M, Deb K, Rudolph G, Yao X, Lutton E, Merelo JJ, Schwefel HP (eds.). Springer: Berlin, Germany, 2000; Volume 1917, pp. 849–858,2000.

[245] Hole MK, Gulhane VS and Shellokar ND. Application of Genetic Algorithm for Image Enhancement and Segmentation. International Journal of Advanced Research in Computer Engineering & Technology (IJARCET). 2013 Apr 28; 2(4): pp-1342.

[246] Mohanta RK and Sethi B. A Study on Application of Artificial Neural Network and Genetic Algorithm in Pattern Recognition. International Journal of Computer Science & Engineering Technology (IJCSET) Vol. 3.

[247] Fan SK and Chang JM. A parallel particle swarm optimization algorithm for multi-objective optimization problems. Engineering Optimization. 2009 Jul 1; 41(7): 673–97.

[248] Gunawan S, Farhang-Mehr A and Azarm S. On maximizing solution diversity in a multiobjective multidisciplinary genetic algorithm for design optimization.

[249] Zhou A, Jin Y, Zhang Q, Sendhoff B and Tsang E. Combining model-based and genetics-based offspring generation for multi-objective optimization using a convergence criterion. In Evolutionary Computation, 2006. CEC 2006. IEEE Congress on 2006 Jul 16 (pp. 892–899). IEEE.

[250] Marghany M. Multi-objective entropy evolutionary algorithm for marine oil spill detection using cosmo-skymed satellite data. Ocean Science Discussions. 2015 May 1; 12(3).

[251] Marghany M and Hakami M. Automatic Detection of Coral Reef Induced Turbulent Boundary Flow in the Red Sea from Flock-1 Satellite Data. In Oceanographic and Biological Aspects of the Red Sea 2019 (pp. 105–122). Springer, Cham.

[252] Zhang B, Perrie W, Li X and Pichel WG. Mapping sea surface oil slicks using RADARSAT-2 quad-polarization SAR image. Geophysical Research Letters. 2011 May 1; 38(10).

[253] Souyris JC, Imbo P, Fjortoft R, Mingot S, Lee JS. Compact polarimetry based on symmetry properties of geophysical media: The/spl pi//4 mode. IEEE Transactions on Geoscience and Remote Sensing. 2005 Mar; 43(3): 634–46.

[254] Migliaccio M, Nunziata F, Brown CE, Holt B, Li X, Pichel W, Shimada M. Polarimetric synthetic aperture radar utilized to track oil spills. Eos, Transactions American Geophysical Union. 2012 Apr 17; 93(16): 161–2.

[255] Kudryavtsev VN, Chapron B, Myasoedov AG, Collard F and Johannessen JA. On dual co-polarized SAR measurements of the ocean surface. IEEE Geoscience and Remote Sensing Letters. 2013 Jul; 10(4): 761–5.

[256] Zheng H, Khenchaf A, Wang Y, Ghanmi H, Zhang Y and Zhao C. Sea surface monostatic and bistatic EM scattering using SSA-1 and UAVSAR data: Numerical evaluation and comparison using different sea spectra. Remote Sensing. 2018 Jul; 10(7): 1084.

[257] Leifer I, Lehr WJ, Simecek-Beatty D, Bradley E, Clark R, Dennison P, Hu Y, Matheson S, Jones CE, Holt B and Reif M. State of the art satellite and airborne marine oil spill remote sensing: Application to the BP Deepwater Horizon oil spill. Remote Sensing of Environment. 2012 Sep 1; 124: 185–209.

[258] De Castro LN and Von Zuben FJ. The clonal selection algorithm with engineering applications. In Proceedings of GECCO 2000 Jul 8 (Vol. 2000, pp. 36–39).

[259] Han KH and Kim JH. Quantum-inspired evolutionary algorithms with a new termination criterion, H/sub/spl epsi//gate, and two-phase scheme. IEEE transactions on evolutionary computation. 2004 Apr;8(2): 156–69.

[260] Jiao L, Li Y, Gong M and Zhang X. Quantum-inspired immune clonal algorithm for global optimization. IEEE Transactions on Systems, Man, and Cybernetics, Part B (Cybernetics). 2008 Oct; 38(5): 1234–53.

[261] Marghany M and van Genderen J. Entropy algorithm for automatic detection of oil spill from radarsat-2 SAR data. In IOP Conference Series: Earth and Environmental Science 2014 (Vol. 18, No. 1, p. 012051). IOP Publishing.

[262] Skrunes S, Brekke C, Eltoft T and Kudryavtsev V. Comparing near-coincident C-and X-band SAR acquisitions of marine oil spills. IEEE Transactions on Geoscience and Remote Sensing. 2015 Apr; 53(4):1958–75.

[263] Matkan AA, Hajeb M and Azarakhsh Z. Oil spill detection from SAR image using SVM based classification. International Archives of the Photogrammetry, Remote Sensing and Spatial Information Sciences, SMPR. 2013 Sep; 1: W3.

[264] Marghany M. Oil Spill Detection from Cosmo-Skymed Satellite Data Multi-Objective Evolutionary Algorithm. International Journal of Petroleum and Petrochemical Engineering (IJPPE), 2018, 4(3): 43–48, DOI: http://dx.doi.org/10.20431/2454-7980.0403005.

[265] Balicki J. An adaptive quantum-based multiobjective evolutionary algorithm for efficient task assignment in distributed systems. In WSEAS International Conference. Proceedings. Recent Advances in Computer Engineering 2009 Jul 23 (No. 13). WSEAS.

[266] Zheng Q, Yan XH, Liu WT, Klemas V, Sun D. Space shuttle observations of open ocean oil slicks. Remote sensing of environment. 2001 Apr 1; 76(1): 49–56.

[267] Cheng Y, Liu B, Li X, Nunziata F, Xu Q, Ding X, Migliaccio M, Pichel WG. Monitoring of oil spill trajectories with COSMO-SkyMed X-band SAR images and model simulation. IEEE Journal of Selected Topics in Applied Earth Observations and Remote Sensing. 2014 Jul; 7(7): 2895–901.

[268] Afenyo M, Veitch B and Khan F. A state-of-the-art review of fate and transport of oil spills in open and ice-covered water. Ocean Engineering. 2016 Jun 1; 119: 233–48.

[269] Mishra AK and Kumar GS. Weathering of oil spill: modeling and analysis. Aquatic Procedia. 2015 Jan 1; 4: 435–42.

[270] Galt JA. Uncertainty analysis related to oil spill modeling. Spill Science & Technology Bulletin. 1997 Jan 1; 4(4): 231–8.

[271] Beegle-Krause J. General NOAA oil modeling environment (GNOME): a new spill trajectory model. In International Oil Spill Conference 2001 Mar (Vol. 2001, No. 2, pp. 865–871). American Petroleum Institute.

[272] Wirtz KW, Baumberger N, Adam S and Liu X. Oil spill impact minimization under uncertainty: Evaluating contingency simulations of the Prestige accident. Ecological Economics. 2007 Mar 1;61(2-3): 417–28.

[273] Sebastião P and Soares CG. Uncertainty in predictions of oil spill trajectories in open sea. Ocean Engineering. 2007 Mar 1; 34(3-4): 576–84.

[274] Fay JA. The spread of oil slicks on a calm sea. In Hoult DP (Ed.), Oil on the Sea, Plenum Press, New York (1969), pp. 53–63.

[275] Cheng Y, Li X, Xu Q, Garcia-Pineda O, Andersen OB and Pichel WG. SAR observation and model tracking of an oil spill event in coastal waters. Marine pollution bulletin. 2011 Feb 1; 62(2): 350–63.

[276] Yapa PD, Shen HT and Angammana KS. Modeling oil spills in a river—lake system. Journal of Marine Systems. 1994 Mar 1; 4(6): 453–71.

[277] Yapa PD and Tao Shen H. Modelling river oil spills: a review. Journal of Hydraulic Research. 1994 Sep 1; 32(5): 765–82.

[278] Marghany M. RADARSAT for oil spill trajectory model. Environmental Modelling & Software. 2004 May 1; 19(5): 473–83.

[278] Spaulding ML. A state-of-the-art review of oil spill trajectory and fate modeling. Oil and Chemical Pollution. 1988 Jan 1; 4(1): 39–55.

[279] Arik S. A note on the global stability of dynamical neural networks. IEEE Transactions on Circuits and Systems I: Fundamental Theory and Applications. 2002 Apr; 49(4): 502–4.

[280] Zhao H. Global asymptotic stability of Hopfield neural network involving distributed delays. Neural networks. 2004 Jan 1; 17(1): 47–53.

[281] Cao J and Wang J. Global asymptotic stability of a general class of recurrent neural networks with time-varying delays. IEEE Transactions on Circuits and Systems I: Fundamental Theory and Applications. 2003 Jan; 50(1): 34–44.

[282] Cote S and Tatnall AR. The Hopfield neural network as a tool for feature tracking and recognition from satellite sensor images. International Journal of Remote Sensing. 1997 Mar 1; 18(4): 871–85.

[283] Nasrabadi NM and Choo CY. Hopfield network for stereo vision correspondence. IEEE transactions on neural networks. 1992 Jan; 3(1): 5–13.

[284] Yi Z, Heng PA and Fu AW. Estimate of exponential convergence rate and exponential stability for neural networks. IEEE Transactions on Neural Networks. 1999 Nov; 10(6): 1487–93.

[285] Wang L, Zhang Y and Zhang Y. On absolute stability for a class of nonlinear control systems with delay. Chinese Science Bulletin. 1993 Mar; 38(16): 1445–1448.

[286] Juang JC. Stability analysis of Hopfield-type neural networks. IEEE Transactions on Neural Networks. 1999 Nov; 10(6): 1366–74.

[287] Arik S. A note on the global stability of dynamical neural networks. IEEE Transactions on Circuits and Systems I: Fundamental Theory and Applications. 2002 Apr; 49(4): 502–4.

[288] Marghany M. Simulation sea surface current from RADARSAT-2 SAR data using Hopfield neural network. In 2015 IEEE 5th Asia-Pacific Conference on Synthetic Aperture Radar (APSAR) 2015 Sep 1 (pp. 805–808). IEEE.

[289] Marghany M. Utilization of Hopfield neural network and quasi-linear model for longshore current pattern simulation from RADARSAT. In IGARSS 2003. 2003 IEEE International Geoscience and Remote Sensing Symposium. Proceedings (IEEE Cat. No. 03CH37477) 2003 Jul 21 (Vol. 4, pp. 2688–2690). IEEE.

[290] Marghany M. Hopfield neural network and pareto optimal algorithms for retrieving sea surface current from Tan DEM-X data. InIOP Conference Series: Earth and Environmental Science 2018 Jun (Vol. 169, No. 1, p. 012023). IOP Publishing.

[291] Behrman EC, Nash LR and Steck JE. Chandrashekar VG, Skinner SR. Simulations of quantum neural networks. Information Sciences. 2000 Oct 1; 128(3-4): 257–69.

[292] Shukla A and Vedula P. Trajectory optimization using quantum computing. Journal of Global Optimization. 2019: 1-27.

[293] Von Stryk O and Bulirsch R. Direct and indirect methods for trajectory optimization. Annals of operations research. 1992 Dec 1; 37(1): 357–73.

[294] Bulger DW. Combining a local search and Grover's algorithm in black-box global optimization. Journal of optimization theory and applications. 2007 Jun 1; 133(3): 289–301.

[295] Dietrich JC, Trahan CJ, Howard MT, Fleming JG, Weaver RJ, Tanaka S, Yu L, Luettich Jr RA, Dawson CN, Westerink JJ and Wells G. Surface trajectories of oil transport along the Northern Coastline of the Gulf of Mexico. Continental Shelf Research. 2012 Jun 1; 41:17–47.

[296] Nunziata F, Buono A and Migliaccio M. COSMO–SkyMed Synthetic Aperture Radar Data to Observe the Deepwater Horizon Oil Spill. Sustainability. 2018 Oct 10; 10(10): 3599.

[297] Walker ND, Pilley CT, Raghunathan VV, D'Sa EJ, Leben RR, Hoffmann NG, Brickley PJ, Coholan PD, Sharma N, Graber HC and Turner RE. Impacts of Loop Current frontal cyclonic eddies and wind forcing on the 2010 Gulf of Mexico oil spill. Monitoring and Modeling the Deepwater Horizon Oil Spill: A Record-Breaking Enterprise, Geophys. Monogr. Ser. 2011 Jan 1; 195: 103–16.

[298] Liu Y, Weisberg RH, Hu C, Kovach C and Riethmüller R. Evolution of the Loop Current system during the Deepwater Horizon oil spill event as observed with drifters and satellites. Monitoring and Modeling the Deepwater Horizon Oil Spill: A Record-Breaking Enterprise, Geophys. Monogr. Ser. 2011 Jan 1; 195: 91–101.

[299] Liu Y, Weisberg RH, Hu C and Zheng L. Satellites, models combine to track Deepwater Horizon oil spill. SPIE Newsroom, doi. 2011; 10(2.1201104): 003575.

[300] Mezić I, Loire S, Fonoberov VA and Hogan P. A new mixing diagnostic and Gulf oil spill movement. Science. 2010 Oct 22; 330(6003): 486–9.

Index

D

de Broglie theory 18
Decoherence 138–140, 157–159, 202, 203, 225, 237, 241, 242
Deepwater Horizon (DWH) 249, 262
Density 2, 10, 11, 24, 43, 46, 51, 52, 72, 73, 80, 135–137, 139, 140, 155–158, 160, 164, 207, 267
Dielectric 3, 5, 8, 10, 60, 61, 128, 131, 134, 136, 158, 223
Dimension 20, 51, 52, 82, 103, 109, 111, 146, 178–182, 184, 185, 187, 192–199, 228
Dirac equation 19–21
Direction 4, 6, 7, 10, 11, 15, 21, 30, 31, 40, 42, 51, 52, 54–56, 59, 61, 65, 80, 81, 83–86, 88, 98, 106, 113, 114, 116, 122, 130–132, 143, 145, 158, 171, 172, 190, 208, 230, 246, 249–251
Discrete 15, 16, 42, 146, 184, 270
Doppler 59, 88–92, 94, 97, 101–103, 112–114, 116, 121, 246, 250
Dual 16, 19, 20, 28, 30, 36, 38, 44, 45, 49, 51, 58, 63, 75, 76, 78, 79, 101, 110, 112, 114, 121, 123, 132, 144, 158, 159, 171, 202, 203, 207, 220, 223–226, 233, 241, 261, 263, 269
Duality 18, 125, 127, 138

E

Eigenvalue 41, 140, 242, 258
Electricity 5, 17, 32
Electrodes 5
Electromagnetic 1, 4–13, 15–21, 23, 25, 27, 28, 31–35, 37, 40, 42, 44–46, 59, 63–66, 68, 69, 72, 74, 75, 80, 81, 108, 117, 127, 129, 138, 139, 142, 221
Electron 16–19, 21–23, 25–27, 28, 31, 38, 51, 52, 60, 61, 75, 76, 126, 128–131, 133, 136, 201
Elliptical 250
Energy 2, 9–12, 15–21, 23, 24, 28, 31–34, 36–38, 41–43, 45, 46, 48, 50–52, 55, 56, 59–63, 65, 66, 68, 72, 75, 80, 88, 108, 113, 114, 125–131, 135–137, 147, 148, 152–155, 235, 252, 254, 255, 257, 258, 270
Entanglement 69 ,76, 77, 158, 207, 220, 256
ENVISAT 106, 160, 261–264, 267
ERS-1/2 82, 106, 107
FSA 106, 261
Euler equation 254
Exponentials 6

F

Fay's algorithm 247–249
Feature 97, 106, 160, 161, 163, 166, 169, 170, 173, 180, 185, 192, 197, 199, 203, 220, 225, 229, 253–258

Fermions 126, 267
Feynman 25–29, 31, 44, 50, 51, 62, 267
Field 15, 17, 52, 56, 60–64, 66, 68, 74, 75, 79, 80, 104, 109, 112, 115, 121, 125, 131, 132, 135, 141, 142, 190, 221, 252, 267, 271
Finite-dimensional 43
Force Carriers 19
Forecasting 14, 20, 247, 248, 250, 267
Four-dimensional 20
Fourier transform 31, 41, 56, 94, 101
Fractal 143, 178–187, 189, 192, 194–199, 220, 245
Frequency 4, 15–17, 32, 33, 35–38, 40, 42, 46, 47, 49, 61, 66, 68, 69, 74, 75, 78, 84, 86, 88–92, 94–98, 101, 102, 106–110, 113–116, 121, 127, 130, 135–137, 142, 143, 145, 149, 150, 153, 161, 190, 200, 208, 233, 246, 250, 261
Function 4, 10, 12, 17, 19, 22–26, 28, 31, 42, 43, 45–48, 50, 52–54, 56, 57, 61, 64, 65, 68, 72, 73, 80, 81, 84, 88–94, 97, 98, 100–102, 106, 108, 116, 121, 125, 126, 128–134, 136–138, 140–143, 152, 160, 164, 168, 170–173, 175, 180, 182–184, 190, 192, 194, 199, 201–204, 207, 215, 220, 222–224, 231, 241–243, 245, 248, 250, 252, 254, 257, 258, 263, 270

G

Gamma wave 33
Gaussian 56, 106
G-band 109
Gear shifts 260
Geometrical optics 31
Geometry 57, 64, 65, 68, 74, 75, 85, 103, 117, 119, 135, 169, 178, 184, 197
Georeference 249
GLCM 143–147, 149, 150, 153, 169
Gradient 130, 136, 169, 172, 258, 270
Ground-range 88
Grover operator 259, 260, 270
Grover search 260, 261, 270
Gulf of Mexico 160, 207, 208, 211, 221, 233, 237, 249, 261–263, 267, 269, 270

H

Hadamard transform 259
Hamiltonian 25, 29, 45, 48, 129, 130, 156, 257, 258
Harmonic oscillators 41
Hebbian 256, 263
Hermite polynomial 25
Hermitian operator 41, 42
Hertz 5, 33, 68
Hilbert space 31, 43, 44, 139, 156, 204, 205, 224
Hypervolume 267

Color Plate Section

Chapter 6

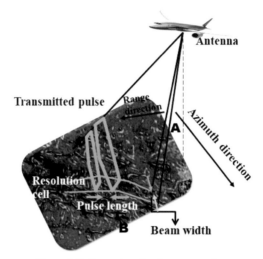

Figure 6.8. Geometry of real aperture radar.

Chapter 9

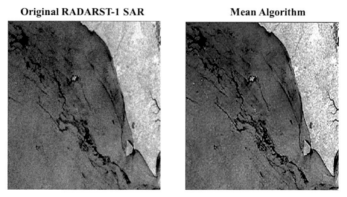

Figure 9.9. GLMC based Mean algorithm.

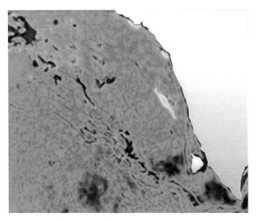

Oil Spill   Look-alikes   Sea surface

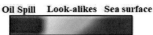

Figure 9.14. Quantum entropy for classification of sea surface features from RADARSAT-1 SAR wide mode acquired on October 26, 1997.

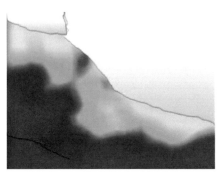

Oil Spill   Look-alikes   Sea surface

Figure 9.15. Quantum entropy for classification of sea surface features from RADARSAT-1 SAR acquired on December 20, 1999.

Chapter 10

Figure 10.12. Neural networks for automatic detection of land and sea surface from S2 data.

Chapter 13

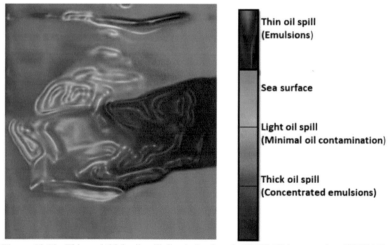

Figure 13.10. Thin and thick oil spill discriminations in UAVSAR image using QNSGAII.

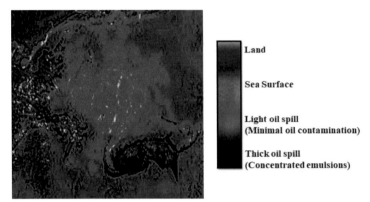

Figure 13.15. QNSGA-II clustering for RADARSAT-2 acquired on April 27th, 2010.

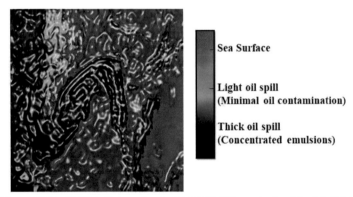

Figure 13.17. QNSGA-II clustering for RADARSAT-2 acquired on May 5th, 2010.

Chapter 14

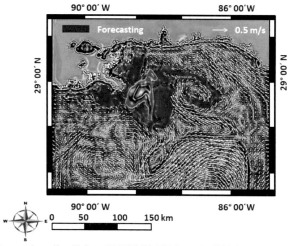

Figure 14.15. Forecasting oil spill from ENVISAT ASAR on April 29th, 2010 to May 10th, 2010 using the qHop algorithm.

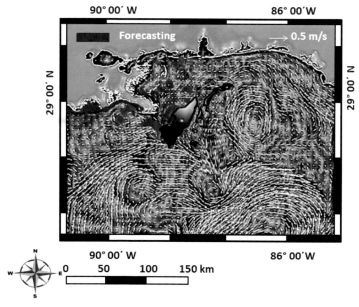

Figure 14.17. Simulation of the oil spill trajectory movement based on CSK SAR on July 20th, 2010.

Printed and bound by CPI Group (UK) Ltd, Croydon, CR0 4YY

17/10/2024

01775667-0003